Cambridge IGCSE™

Biology
STUDY AND REVISION GUIDE

Dave Hayward

HODDER
EDUCATION
AN HACHETTE UK COMPANY

Hachette UK's policy is to use papers that are natural, renewable and recyclable products and made from wood grown in well-managed forests and other controlled sources. The logging and manufacturing processes are expected to conform to the environmental regulations of the country of origin.

Orders: please contact Hachette UK Distribution, Hely Hutchinson Centre, Milton Road, Didcot, Oxfordshire, OX11 7HH. Telephone: +44 (0)1235 827827. Email education@hachette.co.uk. Lines are open from 9 a.m. to 5 p.m., Monday to Friday. You can also order through our website: www.hoddereducation.com

ISBN: 978 1 4718 6513 8

The questions, example answers, marks awarded and/or comments that appear in this book were written by the authors. In an examination, the way marks would be awarded to answers like these may be different. This text has not been through the Cambridge International endorsement process.

© Dave Hayward 2016

First published in 2005

This edition published in 2016 by
Hodder Education,
An Hachette UK Company
Carmelite House
50 Victoria Embankment
London EC4Y 0DZ

www.hoddereducation.com

Impression number 10 9 8

Year   2023

Cover photo © cyberstock / Alamy
Illustrations by Integra Software Services
Typeset by Integra Software Services Pvt. Ltd., India
Printed and bound by CPI Group (UK) Ltd, Croydn, CR0 4YY
A catalogue record for this title is available from the British Library.

Contents

REVISED

Introduction

Welcome to the Cambridge IGCSE™ Biology Study and Revision Guide. This book has been written to help you revise everything you need to know for your Biology exam. Following the Biology syllabus, it covers all the key core and extended content and provides sample questions and answers, as well as practice questions, to help you learn how to answer questions and to check your understanding.

How to use this book

Key objectives

The key skills and knowledge covered in the chapter. You can also use this as a checklist to track your progress.

Key terms

Definitions of key terms you need to know from the syllabus.

Common misconceptions

Mistakes students often make, and how to avoid them.

Sample questions

Exam-style questions for you to think about.

Student's answers

Typical student answers to see how the question might have been answered.

Examiner's comments

Feedback from an examiner showing what was good, and what could be improved.

Now try this

Practice questions for you to answer so that you can see what you have learned.

Examiner's tips

Advice to help you give the perfect answer.

Extended

Content for the extended syllabus is shaded green.

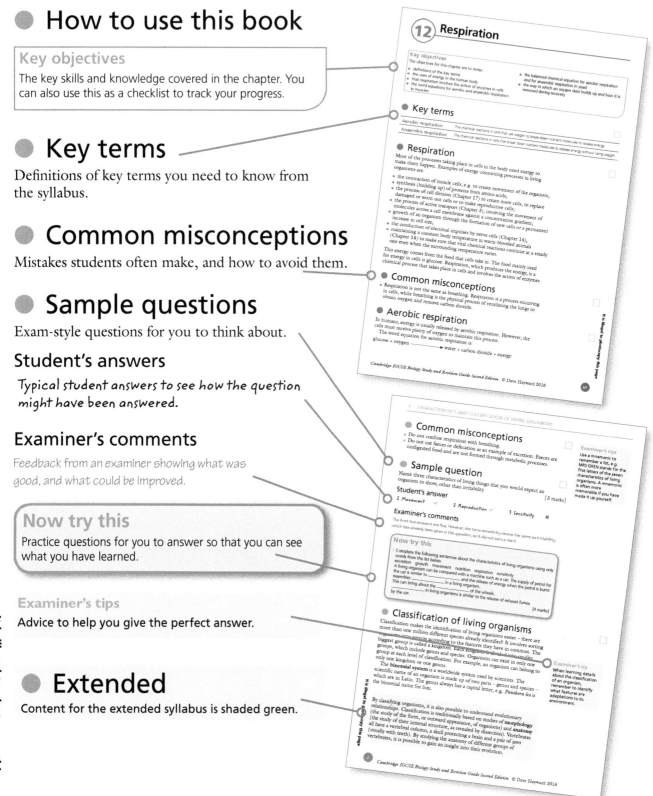

Cambridge IGCSE Biology Study and Revision Guide Second Edition © Dave Hayward 2016

How should I revise?

Studying on your own should be organised so that you are also active. Simply reading through a textbook and hoping you will learn just does not work.

We only really learn things thoroughly when we actually do things with the material – active learning. Not only is this more interesting and motivating, but we actually understand, learn and remember things much better this way.

This book contains a number of sample and practice questions. When reading the sample questions, think about how you would answer the question before looking at the answers. Think about how you would mark the students answer. You can also make notes on the page to help you remember the information as you work through the book.

Theory papers

Once the examination has started, browse through the paper and choose a question you feel confident about. You do not have to start with question 1. Read the question twice, look at the mark allocation for each part and then decide exactly what is required to achieve marks. Never forget – marks are not awarded for correct Biology, but for correct Biology that answers the question. Taking some time to plan your answer carefully can help with this. If you finish early, use the time to check through your answers. Ask yourself, 'Have I answered the question and have I made sufficient points to be awarded all the marks?'

How to answer different types of questions

Calculations

Always show your working. The first reason for this is that if you write down all your working, you are far more likely to work in the logical way needed to reach correct answers. You can also gain marks for using the right method, even if you get part of the answer wrong.

Set your work out logically. Neatness helps logical working and it is important to develop this habit throughout the course. Any calculations you do should be logically presented, and show all the steps of working.

Show the units. Sometimes the question will make it easy for you and give the units. Sometimes you will be asked to state the units. Marks are often given for units, so develop the habit of thinking about the units and writing them down even when not asked for. Scientific quantities are meaningless without units. For example, it makes a lot of difference whether a volume is cm^3 or dm^3

Graphs

Drawing graphs. The axes should be labelled with units. The scales should fill more than half the space available in each direction. Think carefully about whether the origin should or should not be included. There is no need to fill every part of the available space so do not use scales with awkward multiples such as 3 or 7. The dependent variable is always on the y-axis and the independent variable is always on the x-axis. Points should be plotted carefully to an accuracy of 0.5 mm. This means using a sharp pencil to mark the point with a small cross or a circled dot. If the line of best fit is a straight line, carefully judge its position and draw one thin line with a ruler. If the line is a curve, it should be a single, thin, smooth line through the majority of the points. It must not be distorted to pass through every point.

Reading off graphs. Again, you should work to an accuracy of 0.5 mm. Draw vertical and horizontal lines to the axes to show your working.

Descriptions and logical deduction

Logical thinking. Many questions require step-by-step descriptions and/or deductions. You must be just as logical with words as you would be in working out a question with numbers.

How much do I write? The value of the answer is always given in brackets next to the question. So, for example, a description with a mark of [3 marks] will require you to give three valid statements. There are only a certain number of marks given for each question and if you write too much you will not be given any more marks and there is a chance that you will introduce contradictory statements. These can cancel out marks already achieved. The space available is also a rough guide, but not a fixed rule. If you have written too much or too little, think about whether you have included something you don't need, or if you have missed out something important.

Cloze passages

Missing words questions. Read through the paragraph first, then fill in the easiest answers. Cross off the words you use from the list. This will leave you with a smaller choice for the hardest answers.

Matching pairs

Matching names with descriptions, definitions or functions. You are sometimes required to draw a line between the name and its function. When answering this type of question use a ruler – curved lines can be ambiguous – and draw the line with a pencil (then, if you make a mistake, you can correct it).

● Multiple choice papers

For all the questions on this paper you will be given a choice of answers. When unsure of the answer, try to eliminate some of the incorrect options as this increases the chance of choosing the correct response, and remember that there will often be a 'distractor' – an incorrect answer that might seem correct. If you cannot answer a question, put a mark by it in the margin of the paper, leave it, and return to it when you have completed the other questions. If you still can't answer the question, it is always worth having a guess as you will not be penalised for wrong answers, but this should only be done as a last resort.

● Examination terms explained

The examination syllabus gives a full list of the terms used by examiners and how candidates are expected to respond.

Calculate	Give a numerical answer, generally showing the working out involved
Define	A precise statement is needed
Describe	State the main details, without an explanation
Discuss	Involves giving different points of view, or advantages and disadvantages
Explain	You must give reasons and/or underlying theory
List	You must give a series of points (usually as single words), keeping to the number specified in the question
Outline	State the main facts briefly, without going into detail
Predict	You are not supposed to know the answer from memory, but to deduce it, usually from information in the question
State	Give a concise answer; no explanation is needed
Suggest	This implies there is more than one acceptable answer or that candidates are expected to arrive at the answer using their general knowledge of Biology
What do you understand by	Give the definition and some additional explanation

Cambridge IGCSE Biology Study and Revision Guide Second Edition © Dave Hayward 2016

● Practical examinations

There are two ways of assessing practical skills: the practical test or the written test of practical skills.

The written test is a single written paper devoted entirely to laboratory procedures. The syllabus gives a complete list of the required procedures. The practical test will present different problems from a purely written assessment – in a word, nerves. In any examination you need to be calm and measured in your approach. In a practical examination, think about what you are about to do, and when you are certain of the correct action, carry it out. Do not rush. If you make a mistake, you might have to start the exercise again.

● How to improve your grade

Here are a few brief summary points, all of which have been mentioned elsewhere in this book.

- Use this book – it has been written to help students achieve high grades.
- Learn all the work – low grades are nearly always attributable to inadequate preparation. If you can recall the work, you will succeed. Don't leave things to chance.
- Practise skills such as calculations, equation writing, labelling diagrams and the interpretation of graphs.
- Use past papers to reinforce revision, to become familiar with the type of question and to gain confidence.
- Answer the question as instructed on the paper – be guided by the key words used in the question (describe, explain, list, etc.). Do not accept a question as an invitation to write about the topic.

Characteristics and classification of living organisms

1

The objectives for this chapter are to revise:

- definitions of the key terms
- characteristics of living organisms
- classification of organisms into groups using shared features
- species and the binomial system of naming them
- features in the cells of all living organisms
- features used to place animals and plants into the appropriate kingdoms

- classification of vertebrates and arthropods
- the construction and use of simple dichotomous keys
- the use of morphology, anatomy and cladistics in classification
- the five kingdoms
- classification of flowering plants and viruses

● Key terms

Nutrition	The taking in of materials for energy, growth and development
Excretion	The removal from organisms of toxic materials and substances that are in excess of requirements
Respiration	The chemical reactions in cells that break down nutrient molecules and release energy
Sensitivity	The ability to detect and respond to changes in the environment
Reproduction	The processes that make more of the same kind of organism
Growth	A permanent increase in size
Movement	An action by an organism causing a change of position or place
Species	A group of organisms that can reproduce to produce fertile offspring
Binomial system	An internationally agreed system in which the scientific name of an organism is made up of two parts: the genus and the species
Nutrition	The taking in of materials for energy, growth and development. Plants require light, carbon dioxide, water and ions. Animals need organic compounds and ions and usually need water
Excretion	The removal from organisms of the waste products of metabolism (chemical reactions in cells including respiration), toxic materials and substances that are in excess of requirements
Respiration	The chemical reactions in cells that break down nutrient molecules and release energy for metabolism
Sensitivity	The ability to detect or sense stimuli in the internal or external environment and to respond appropriately
Growth	A permanent increase in size and dry mass by an increase in cell number or cell size, or both
Movement	An action by an organism or part of an organism causing a change of position or place

● Characteristics of living organisms

There are seven characteristics that all living things show, including plants and other organisms. These characteristics are movement, respiration, sensitivity, growth, reproduction, excretion and nutrition. You need to be able to recall and describe these. You may be given a picture of a living thing to study and then asked to identify which characteristics you can observe by watching it for a few minutes. Some of the seven would not be suitable answers, e.g. growth, respiration, reproduction (these are not likely to be visible or observable in a short time span). Some non-living things, such as cars, may appear to show some of the characteristics, but not all of them.

● Common misconceptions

- Do not confuse respiration with breathing.
- Do not use faeces or defecation as an example of excretion. Faeces are undigested food and are not formed through metabolic processes.

● Sample question

Name three characteristics of living things that you would expect an organism to show, other than irritability.   [3 marks]

Student's answer

1 Movement   ✓   2 Reproduction ✓   3 Sensitivity   ✗

Examiner's comments

The first two answers are fine. However, the term sensitivity means the same as irritability, which has already been given in the question, so it did not earn a mark.

> **Now try this**
>
> 1 Complete the following sentences about the characteristics of living organisms using only words from the list below.
> excretion   growth   movement   nutrition   respiration   sensitivity
> A living organism can be compared with a machine such as a car. The supply of petrol for the car is similar to _____ and the release of energy when the petrol is burnt resembles _____ in a living organism.
> This can bring about the _____ of the wheels.
> _____ in living organisms is similar to the release of exhaust fumes by the car.   [4 marks]

Examiner's tips

Use a mnemonic to remember a list, e.g. MRS GREN stands for the first letters of the seven characteristics of living organisms. A mnemonic is often more memorable if you have made it up yourself.

● Classification of living organisms

Classification makes the identification of living organisms easier – there are more than one million different species already identified! It involves sorting organisms into groups according to the features they have in common. The biggest group is called a kingdom. Each kingdom is divided into smaller groups, which include genus and species. Organisms can exist in only one group at each level of classification. For example, an organism can belong to only one kingdom or one genus.

The **binomial system** is a worldwide system used by scientists. The scientific name of an organism is made up of two parts – genus and species – which are in Latin. The genus always has a capital letter, e.g. *Panthera leo* is the binomial name for lion.

Examiner's tip

When learning details about the classification of an organism, remember to identify what features are adaptations to its environment.

By classifying organisms, it is also possible to understand evolutionary relationships. Classification is traditionally based on studies of **morphology** (the study of the form, or outward appearance, of organisms) and **anatomy** (the study of their internal structure, as revealed by dissection). Vertebrates all have a vertebral column, a skull protecting a brain and a pair of jaws (usually with teeth). By studying the anatomy of different groups of vertebrates, it is possible to gain an insight into their evolution.

Cambridge IGCSE Biology Study and Revision Guide Second Edition   © Dave Hayward 2016

The sequences of DNA and of amino acids in proteins are used as a more accurate means of classification than studying morphology and anatomy. Eukaryotic organisms contain chromosomes, made up of strings of genes. Genes are made of DNA, which is composed of a sequence of bases (see Chapter 4). Each species has a distinct number of chromosomes and a unique sequence of bases in its DNA, making it identifiable and distinguishable from other species.

The process of biological classification called **cladistics** involves organisms being grouped together according to whether or not they have one or more unique characteristics in common derived from the group's last common ancestor, which are not present in more distant ancestors. Organisms that share a more recent ancestor (and are more closely related) have DNA base sequences that are more similar than those that share only a distant ancestor.

● Features of organisms

The cells of all living organisms contain cytoplasm, a cell membrane and DNA as genetic material. Two kingdoms are the plant and animal kingdoms.

Plants are made up of many cells – they are multi-cellular. Plant cells have an outside wall made of cellulose. Many of the cells in plant leaves and stems contain chloroplasts with photosynthetic pigments, e.g. chlorophyll. Plants make their food through photosynthesis.

Animals are multi-cellular organisms whose cells have no cell walls or chloroplasts. Most animals ingest solid food and digest it internally.

For the Core syllabus, you only need to learn the main groups of vertebrates and arthropods.

● Classification of vertebrates

Vertebrates are animals with backbones (part of an internal skeleton). Vertebrates are divided into five groups called classes. Details of each group are given in the table below. You only need to be able to describe visible external features, but other details can be helpful (see the 'Other details' column).

Vertebrate class	Body covering	Movement	Reproduction	Sense organs	Other details	Examples
Fish	Scales	Fins (also used for balance)	Usually produces jelly-covered eggs in water	Eyes but no ears; lateral line along body for detecting vibrations in water	Cold-blooded; gills for breathing	Herring, perch, shark
Amphibians	Moist skin	Four limbs; back feet often webbed to make swimming more efficient	Produces jelly-covered eggs in water	Eyes and ears	Cold-blooded; lungs and skin for breathing	Frog, toad, salamander
Reptiles	Dry, with scales	Four legs (apart from snakes)	Eggs with rubbery, waterproof shell; eggs are laid on land	Eyes and ears	Cold-blooded; lungs for breathing	Crocodile, python
Birds	Feathers, scales on legs	Wings; two legs	Eggs with hard shell	Eyes and ears	Warm-blooded; lungs for breathing; beak	Flamingo, pigeon
Mammals	Fur	Four limbs	Live young	Eyes, ears with pinna (external flap)	Warm-blooded; lungs for breathing; females have mammary glands to produce milk to feed young; four types of teeth	Elephant, mouse

Now try this

2 Figure 1.1 can be used to identify the main classes of vertebrate. Use the key to identify the main classes represented by the letters A–E. [5 marks]

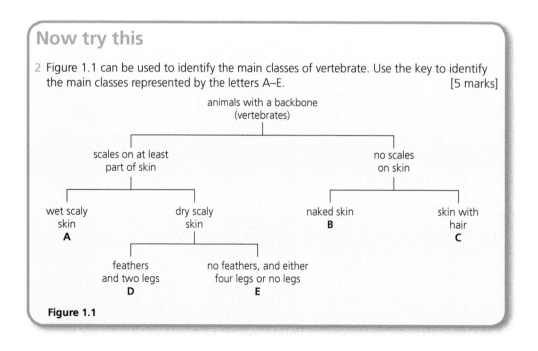

Figure 1.1

Sample question

Animals A, B and C are vertebrates:

- A has a scaly skin, four legs and lungs.
- B has hair, four legs and mammary glands.
- C has a scaly skin, fins and gills.

Create a table to show the group of organisms that each of the animals belongs to. [3 marks]

Student's answer

Animal	Vertebrate group
A	Reptile ✔
B	Mammel ✔
C	Fish ✔

Examiner's comments

The candidate has gained all three marks. The examiner allowed the second answer, although the spelling of mammal was not correct. Try to make sure that your spellings are correct – poor spelling can result in a mark not being awarded, especially if the word is similar to another biological word, e.g. meiosis and mitosis.

Classification of arthropods

Special features of arthropods:

- they are invertebrates – they have no backbone;
- they have an exoskeleton that is waterproof – making arthropods an extremely successful group, as they can exist in very dry places – and they are not confined to water or moist places like most invertebrates;
- their bodies are segmented;
- they have jointed limbs (the exoskeleton would prevent movement).

There are more arthropods than any other group of animals, so they are divided into classes. Figure 1.2 shows the differences between the four classes – insects, arachnids, crustaceans and myriapods. You only need to know about their external features.

Insects, e.g. dragonfly, locust	**Arachnids**, e.g. spider, tick
Key features: • three pairs of legs; • usually have two pairs of wings; • one pair of antennae; • body divided into head, thorax and abdomen; • a pair of compound eyes.	Key features: • four pairs of legs; • body divided into cephalothorax and abdomen; • several pairs of simple eyes; • chelicerae for biting and poisoning prey.
Crustaceans, e.g. crab, woodlouse	**Myriapods**, e.g. centipede, millipede
Key features: • five or more pairs of legs; • two pairs of antennae; • body divided into cephalothorax and abdomen; • exoskeleton often calcified to form a carapace (hard); • compound eyes.	 Key features: • ten or more pairs of legs (usually one pair per segment); • one pair of antennae; • body not obviously divided into thorax and abdomen; • simple eyes.

Figure 1.2

Now try this

3 a Copy the diagrams of the insect, crustacean and arachnid in Figure 1.2 and label the key features that you can see. [4 marks]

b Copy the myriapod diagram in Figure 1.2 and label the features that are common to all arthropods. [3 marks]

● Common misconceptions

● Candidates are often confused by the different numbers of legs in insects, arachnids and crustaceans: candidates often state that insects have three legs instead of three pairs of legs, losing the mark through carelessness or haste. Be careful with your wording!

In addition to the main features of all living organisms, you need to be aware that cells of living organisms also contain **ribosomes** in the cytoplasm, floating freely or attached to membranes called rough endoplasmic reticulum. Ribosomes are responsible for protein synthesis. Enzymes are involved in respiration.

In the classification of living organisms, there are five kingdoms, each with its own special and obvious features. The kingdoms are as follows:

● Animals – multi-cellular organisms that have to obtain their food. Their cells do not have walls.
● Plants – multi-cellular organisms with the ability to make their own food t hrough photosynthesis because of the presence of chlorophyll. Their cells have walls (containing cellulose).
● Fungi – many are made of hyphae, with nuclei and cell walls (containing chitin) but no chloroplasts.
● Prokaryotes (bacteria) – very small and single celled, with cell walls but no nucleus.
● Protoctists – single celled with a nucleus. Some have chloroplasts.

● Features of the plant kingdom

You only need to learn the features of flowering plants and ferns.

Flowering plants are all multi-cellular organisms. Their cells have cellulose cell walls and sap vacuoles. Some of the cells contain chloroplasts.

They have roots, stems and leaves. Reproduction can be by producing seeds, although asexual reproduction is also possible.

There are two groups – monocotyledons and dicotyledons. The term cotyledon means 'seed leaf'. The main differences between the two groups are shown in the table below.

Feature	Monocotyledon	Dicotyledon
Leaf shape	Long and narrow	Broad
Leaf veins	Parallel	Branching
Cotyledons	One	Two
Grouping of flower parts, e.g. petals, sepals and carpels	In threes	In fives

Ferns are land plants. Their stems, leaves and roots are very similar to those of the flowering plants. The stem is usually entirely below ground and takes the form of a structure called a **rhizome**. The stem and leaves have sieve tubes and water-conducting cells. Ferns also have multi-cellular roots with vascular tissue. The leaves are several cells thick. Most of these have an upper and lower epidermis, a layer of palisade cells and a spongy mesophyll. Ferns do not form buds. The midrib and leaflets of the young leaf are tightly coiled and unwind as it grows. Ferns produce gametes but no seeds. The zygote gives rise to the fern plant, which then produces single-celled spores from numerous **sporangia** (spore capsules) on its leaves. The sporangia are formed on the lower side of the leaf.

> **Now try this**
>
> 4 Make your own mnemonic for the five kingdoms with the letters P, P, F, P, A.   [2 marks]

● Features of viruses

Viruses, e.g. HIV

Key features:
- they are very small (100 times smaller than bacteria);
- they do not have a typical cell structure;
- they contain a strand of DNA or RNA;
- they are surrounded by a protein coat called a capsid;
- the only life process they show is reproduction (inside host cells).

Figure 1.3

Cambridge IGCSE Biology Study and Revision Guide Second Edition © Dave Hayward 2016

Using simple dichotomous keys

Keys are often used by biologists in the process of identifying organisms. You need to be able to construct and use a dichotomous key, i.e. a key that branches into two at each stage, requiring you to choose between alternatives.

You have already used a simple example when answering 'Now try this' question 2 in Chapter 1. The first choice was scales on at least part of the skin or no scales on the skin. The example in 'Now try this' question 5 also uses a simple dichotomous key.

Now try this

5 Figure 1.4 shows single leaves from six different trees.
Use the key below to identify which tree each leaf comes from. Make a table similar to the one below and put a tick in the correct box to show how you identified each leaf. Give the name of the tree. Leaf A has been identified for you as an example.

1	a Leaf with smooth outline	go to 2
	b Leaf with jagged outline	go to 3
2	a Leaf about the same length as width	*Cydonia*
	b Leaf about twice as long as it is wide	*Magnolia*
3	a Leaf divided into more than two distinct parts	go to 4
	b Leaf not divided into more than two distinct parts	go to 5
4	a Leaf divided into five parts	*Aesculus*
	b Leaf divided into ten or more parts	*Fraxinus*
5	a Leaf with pointed spines along its edge	*Ilex*
	b Leaf with rounded lobes along its edge	*Quercus*

[5 marks]

Figure 1.4

Leaf	1a	1b	2a	2b	3a	3b	4a	4b	5a	5b	Name of tree
A	✓		✓								*Cydonia*
B											

● Common misconceptions

● Answers are often wrong for this type of question because the candidate has not worked through the key properly to select the answer, but has jumped to a statement that appears to fit the organism.

● Construction of dichotomous keys

You need to be able to develop the skills to construct simple dichotomous keys, based on easily identifiable features. If you know the main characteristics of a group, it is possible to draw up a systematic plan for identifying an unfamiliar organism. The first question should be based on a feature that will split the group into two. The question is going to generate a 'yes' or 'no' answer. For each of the two subgroups formed, a further question based on the features of some of that subgroup should then be developed. This questioning can be continued until every member of the group has been separated and identified.

Cambridge IGCSE Biology Study and Revision Guide Second Edition © Dave Hayward 2016

Organisation of the organism

Key objectives

The objectives for this chapter are to revise:

- definitions of the key terms
- structures of plant and animal cells and the functions of cell structures
- tissues, organs and organ systems
- calculating the magnification and size of biological specimens

● Key terms

Tissue	A group of cells with similar structures working together to perform a shared function (job)
Organ	A structure made up of a group of tissues working together to perform specific functions
Organ system	A group of organs with related functions working together to perform body functions

● Cell structure and organisation

Most living things are made of cells – microscopic units that act as building blocks. Multi-cellular organisms are made up of many cells. Cell shape varies depending on its function (what job it does). Plant and animal cells differ in size, shape and structure (see Figure 2.1). Plant cells are usually larger than animal cells.

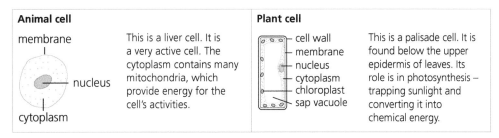

Animal cell

membrane

nucleus

cytoplasm

This is a liver cell. It is a very active cell. The cytoplasm contains many mitochondria, which provide energy for the cell's activities.

Plant cell

- cell wall
- membrane
- nucleus
- cytoplasm
- chloroplast
- sap vacuole

This is a palisade cell. It is found below the upper epidermis of leaves. Its role is in photosynthesis – trapping sunlight and converting it into chemical energy.

Figure 2.1

Examiner's tips

- Practise labelling parts of a plant cell. Start from the outside (the cell wall) and work inwards. This is the correct order: cell wall, membrane, cytoplasm, chloroplast, nucleus, sap vacuole. The chloroplasts and nucleus are both held inside the cytoplasm.
- When labelling a plant cell, make the cell wall label line touch the outer line (cell walls are always drawn as a double line to show their thickness). The membrane label line should touch the inner line of the cell wall (when plant cells are turgid – firm – the membrane is pressed against the cell wall).
- Remember that animal cells contain only three main parts: membrane, nucleus, cytoplasm. Make yourself a mnemonic with MNC or use this one: Mice Nibble Cheese.

Parts of a cell

	Part	Description	Where found	Function
Animal and plant cells	Cytoplasm	Jelly-like, containing particles and organelles	Enclosed by cell membrane	• Contains cell organelles, e.g. mitochondria, nucleus • Chemical reactions take place here
	Membrane	Partially permeable layer that forms a boundary around the cytoplasm	Around the cytoplasm	• Prevents cell contents from escaping • Controls what substances enter and leave the cell
	Nucleus	Round or oval structure containing DNA in the form of chromosomes	Inside the cytoplasm	• Controls cell division • Controls cell development • Controls cell activities
Plant cells	Cell wall	Tough, non-living layer made of cellulose. It surrounds the membrane	Around the outside of plant cells	• Prevents plant cells from bursting • Freely permeable (allows water and salts to pass through)
	Sap vacuole	Fluid-filled space surrounded by a membrane	Inside the cytoplasm of plant cells	• Contains salts and sugars • Helps keep plant cells firm
	Chloroplast	Organelle containing chlorophyll	Inside the cytoplasm of some plant cells	• Traps light energy for photosynthesis

● Common misconceptions

- Remember that animal cells never have a cell wall, chloroplasts or sap vacuoles (although they may have temporary vacuoles where food is stored).
- Remember that not all cells have all cell parts when mature, e.g. red blood cells do not have a nucleus and xylem cells do not have a nucleus or cytoplasm.
- It is not true that all plant cells contain chloroplasts, e.g. epidermis cells and root cells do not.
- Remember that chloroplasts (structures or organelles) are different from chlorophyll (the chemical found in them).

● Sample question

Figure 2.2 shows a nerve cell. State the names of the cell parts A, B and C.   [3 marks]

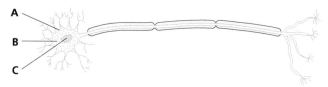

Figure 2.2

Student's answer

A: Cell wall   ✖   B: Cytoplasm   ✔   C: Nucleus   ✔

Examiner's comments

The first answer is wrong – a nerve cell is an animal cell, so it does not have a cell wall. The correct answer for A is cell membrane.

Now try this

1 Trace, copy or sketch the cells shown in Figure 2.1. Practise labelling both cells. Then do the same with other types of animal and plant cells.   [6 marks]

Cambridge IGCSE Biology Study and Revision Guide Second Edition   © Dave Hayward 2016

In addition to the Core material, you need to be able to state that cells also contain other structures in the cytoplasm: ribosomes (found on rough endoplasmic reticulum), mitochondria and vesicles. Only prokaryotes (bacteria) do not have these features. See Figure 2.3.

Ribosomes are responsible for the synthesis of proteins from amino acids (see Chapter 17).

Mitochondria (singular: mitochondrion) are where aerobic respiration takes place. Cells with high rates of metabolism, e.g. liver cells, shown in Figure 2.3, require large numbers of mitochondria to provide sufficient energy.

Vesicles are bubbles of liquid surrounded by a membrane. They transfer chemicals such as enzymes around the cell, or release their contents out of the cell.

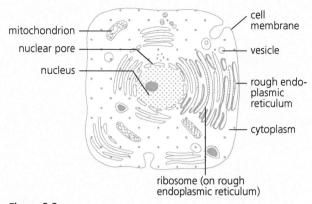

Figure 2.3

Levels of organisation

Figure 2.4 shows examples of cells and their functions in tissues.

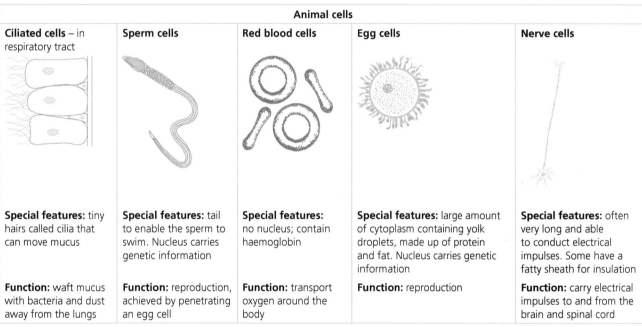

Animal cells				
Ciliated cells – in respiratory tract	**Sperm cells**	**Red blood cells**	**Egg cells**	**Nerve cells**
Special features: tiny hairs called cilia that can move mucus	**Special features:** tail to enable the sperm to swim. Nucleus carries genetic information	**Special features:** no nucleus; contain haemoglobin	**Special features:** large amount of cytoplasm containing yolk droplets, made up of protein and fat. Nucleus carries genetic information	**Special features:** often very long and able to conduct electrical impulses. Some have a fatty sheath for insulation
Function: waft mucus with bacteria and dust away from the lungs	**Function:** reproduction, achieved by penetrating an egg cell	**Function:** transport oxygen around the body	**Function:** reproduction	**Function:** carry electrical impulses to and from the brain and spinal cord

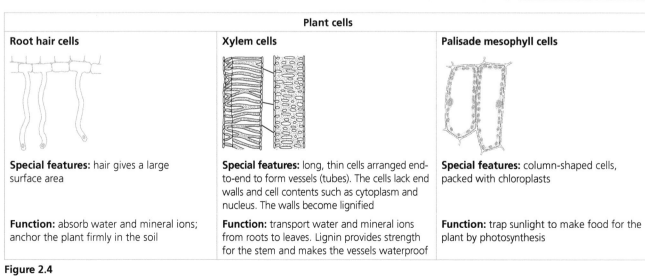

Plant cells		
Root hair cells	**Xylem cells**	**Palisade mesophyll cells**
Special features: hair gives a large surface area	**Special features:** long, thin cells arranged end-to-end to form vessels (tubes). The cells lack end walls and cell contents such as cytoplasm and nucleus. The walls become lignified	**Special features:** column-shaped cells, packed with chloroplasts
Function: absorb water and mineral ions; anchor the plant firmly in the soil	**Function:** transport water and mineral ions from roots to leaves. Lignin provides strength for the stem and makes the vessels waterproof	**Function:** trap sunlight to make food for the plant by photosynthesis

Figure 2.4

Cambridge IGCSE Biology Study and Revision Guide Second Edition © Dave Hayward 2016

● Common misconceptions

- Xylem and phloem tissues are often confused. Xylem carries water and mineral salts, while phloem transports sugars and amino acids.
- Remember that, in a vascular bundle in a stem, phloem is on the outside and xylem is on the inside.

Examiner's tips

- You need to be able to give examples of tissues, organs and organ systems in both plants and animals. A leaf is an organ made up of a number of tissues, e.g. upper epidermis, palisade mesophyll.
- If you draw a diagram to support an exam answer, make sure you refer to it in your written answer. Annotation is more likely to help you gain extra marks.
- It is important that you can identify the different levels of organisation in drawings, diagrams and images of plant and animal material. Practise this by looking at examples in textbooks or on the internet.

● Example of annotation

Figure 2.5 shows the action of a phagocyte.

● Size of specimens

A microscope makes a specimen appear larger than it really is (it magnifies the specimen). You need to be able to calculate the magnification and also the actual size of the specimen.

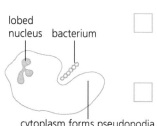

lobed nucleus   bacterium

cytoplasm forms pseudopodia to surround and engulf bacteria – enzymes are released to digest and kill bacteria

Figure 2.5

Examiner's tips

- To calculate the magnification of specimens that have been observed using a light microscope, memorise and use this equation:

$$\text{magnification} = \frac{\text{observed size}}{\text{actual size}}$$

- Make sure the observed size and actual size have the same units.

● Sample question

With reference to a suitable named example, define the term tissue. [3 marks]

Student's answer

A tissue is a group of cells ✔   carrying out the same job. ✔

Examiner's comments

The answer needs three clear points to gain the 3 marks available. This candidate has not named a type of tissue (even though this was the first instruction in the question) and has given only two correct points. Always use the marks shown in the margin to show you how many points to give. Avoid giving more than three; this would waste time that you need to answer other questions. Choose three statements to make before writing them down. The examiner will not select the best answers from a mixture of good and bad ones.

Now try this

2 Identify parts A, B, C and D shown in Figure 2.6 and describe their main features and functions. [12 marks]

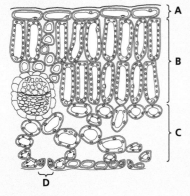

Figure 2.6

3 The diagram of a cow's eye shown in Figure 2.7 is magnified ×2.5 (not drawn to scale). Calculate the actual width of the eye, as shown between points A and B. Show your working. [2 marks]

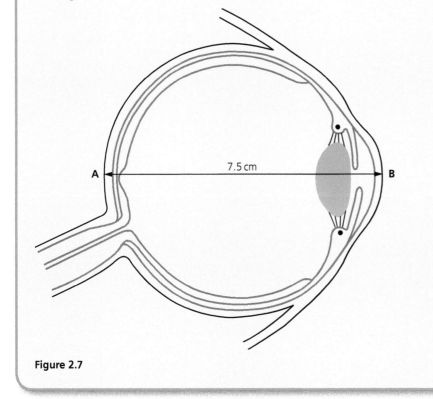

7.5 cm

Figure 2.7

Examiner's tip

When answering the Extended paper, remember there are 1000 micrometres (μm) in a millimetre.

3 Movement in and out of cells

> ## Key objectives
>
> The objectives for this chapter are to revise:
>
> - definitions of the key terms
> - the importance of diffusion of gases and solutes
> - that substances move into and out of cells by diffusion through the cell membrane
> - that water diffuses through partially permeable membranes by osmosis
> - the effects of osmosis on plant and animal tissues
> - the source of energy for diffusion
> - how to describe and explain the effects of factors that influence diffusion
> - how to explain the effects of osmosis on plant tissues
> - how to use the terms associated with osmosis
> - the importance of water potential and osmosis in the uptake of water by plants and its effects on animal cells and tissues
> - how plants are supported by the turgor pressure within cells
> - the importance of active transport and how protein molecules move particles across a membrane during the process

● Key terms

Diffusion	The net movement of particles from a region of their higher concentration to a region of their lower concentration down a concentration gradient, as a result of their random movement
Active transport	The movement of particles through a cell membrane from a region of lower concentration to a region of higher concentration using energy from respiration
Osmosis	The net movement of water molecules from a region of higher water potential (a dilute solution) to a region of lower water potential (a concentrated solution) through a partially permeable membrane

● Diffusion

Diffusion is a really important process for living organisms because it helps to provide essential gases and solutes (materials in solution) and also helps to remove some substances that are potentially toxic (poisonous). These move in or out of the cell through the cell membrane.

Site of diffusion	Substance	Description
Alveoli of lungs	Oxygen	From the alveoli into blood capillaries
Alveoli of lungs	Carbon dioxide	From blood capillaries into the alveoli
Stomata of leaf	Oxygen	From the air spaces, through stomata, into the atmosphere during photosynthesis

The energy for diffusion comes from the kinetic energy (movement energy) of the random movement of molecules and ions. From the organism's point of view, it is a 'free' process – no energy needs to be provided to make it happen.

You need to be able to state the factors that help diffusion to be efficient. These are:

- Distance (the shorter the better), e.g. thin walls of alveoli and capillaries.
- Concentration gradient (the bigger the better). This can be maintained by removing the substance as it passes across the diffusion surface. (Think about oxygenated blood being carried away from the surface of alveoli.)

- Size of the molecules (the smaller the better).
- Surface area for diffusion (the larger the better), e.g. there are millions of alveoli in a lung, giving a huge surface area for the diffusion of oxygen.
- Temperature (molecules have more kinetic energy at higher temperatures).

Common misconceptions

- Do not confuse cell walls with capillary walls – animal cells do not have walls. Many candidates throw away marks referring to 'the thin cell walls of capillaries'. What they mean is 'the *walls of capillaries* are thin because they are only one cell thick'.

Osmosis

Plants rely on osmosis to obtain water through their roots. They use water as a transport medium (to carry mineral salts, sucrose and amino acids around the plant) and to maintain the turgidity of cells (their firmness). When young plants lose more water than they gain, cells become flaccid and the plant wilts.

Fish living in salt water lose water by osmosis. They have very efficient kidneys to reduce water loss in urine.

If we get dehydrated, water is lost from our red blood cells by osmosis. As the cells shrink, they become less efficient at carrying oxygen.

Examiner's tip

Osmosis is a special form of diffusion. It always involves the movement of water across a partially permeable membrane.

Effects of osmosis on plant and animal tissues

- When placed in water, plant and animal cells will take in the water through their cell membranes by diffusion. The diffusion of water in this way is called osmosis.
- Plant cells become turgid (swollen), but do not burst because of their tough cell wall.
- Plants are supported by the pressure of water inside the cells pressing outwards on the cell wall.
- Animal cells will burst, because they have no cell wall.
- The reverse happens when plant and animal cells are placed in concentrated sugar or salt solutions.
- Plant cells become flaccid (limp).
- Animal cells also become flaccid.

Common misconceptions

- Sugars and salts do not move by osmosis. Cell membranes prevent these substances entering or leaving the cell.

Sample question

Some sugar solution was collected from the phloem of a plant stem. Plant cells were placed on a microscope slide and covered with this sugar solution.

Describe what changes would occur to each of the following three cell parts if the sugar solution was more concentrated than the sap in the cell vacuole: sap vacuole, cytoplasm and cell wall. [3 marks]

Student's answer

Sap vacuole: it will get smaller ✓ because there is a higher concentration of water inside the cell, so the water will pass out of the vacuole by osmosis.

Cytoplasm: this will shrink because it is losing water? ✗

Cell wall: this will stop stretching and stop curving outwards. ✓

Examiner's comments

The first answer is correct, but this candidate has wasted time writing more than is needed – the question required a description, not an explanation. The second answer should give details about the way the cytoplasm comes away from the cell wall. In the third answer, details about the cell wall are not very well worded, but it is clear that the candidate understands what is happening.

For the Extended exam, you need to be able to explain the effects of different concentrations of solutions on living tissues:

- When placed in water, plant and animal cells will take in the water through their cell membranes because there is a higher **water potential** outside the cells than inside.
- Plant cells become **turgid** (swollen), but do not burst because of their tough cell wall. A **turgor pressure** is created, which will restrict any further entry of water into the cell.
- Plants are supported by the pressure of water inside the cells pressing outwards on the cell wall, which is inelastic and prevents further net entry of water.
- Animal cells will burst, because they have no cell wall.
- The reverse happens when plant and animal cells are placed in concentrated sugar or salt solutions. This is because there is a higher water potential inside the cell than outside it.
- Plant cells become **flaccid** (limp) and the cytoplasm is no longer pressed against the cell wall. The process of water loss from a cell when it is placed in a solution with a lower water potential is called **plasmolysis**.
- When animal cells become flaccid, their shape can change – they can become crenated.

Examiner's tip

- The term *water potential* means the potential for water to move.
- Water always moves from a higher water potential to a lower water potential.
- A weak (dilute) solution (or pure water) has a high water potential; a strong (concentrated) solution has a low water potential.

Now try this

1 A potato was set up as shown in Figure 3.1 (left-hand side). The investigation was left for several hours. The results are shown on the right-hand side of the figure.

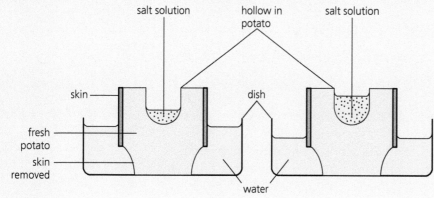

Figure 3.1

a Describe what happened to
   (i) the water in the dish
   (ii) the salt solution in the hollow in the potato. [2 marks]
b (i) Name the process that is responsible for the changes
      that have occurred. [1 mark]
   (ii) Explain why these changes have occurred. [3 marks]
   (iii) Where does this process occur in a plant? [1 mark]
   (iv) What is the importance to the plant of this process? [1 mark]

● Active transport

Active transport is the movement of particles through a cell membrane from a region of lower concentration to a region of higher concentration using energy from respiration.

Animals and plants rely on active transport to move some substances because the concentration gradient is not always the right way round for diffusion. However, cells need to provide energy to achieve movement by active transport. This energy is supplied through respiration using adenosine triphosphate (ATP). Mitochondria (cell organelles in the cytoplasm) control energy release. Respiratory poisons block energy release, so they can prevent active transport.

Protein molecules in the cell membrane play an important part in moving particles across the membrane. The protein binds with ATP, which breaks down to provide energy to move the particles against their concentration gradient.

Examiner's tip

There are two big differences between diffusion and active transport:
● direction of movement (down a gradient, or up a gradient);
● whether or not energy is needed for the movement.

Examples of active transport

Site of active transport	Substance	Description
Root hair cells	Mineral salts, e.g. phosphate	From the soil into the roots
Wall of small intestine (villi)	Glucose	From the small intestine into the blood plasma
Kidney tubules	Glucose	From the filtrate in the tubule into a blood capillary

Now try this

2 Figure 3.2 shows part of a root.
  a Explain how the presence of root hair cells on roots enables the efficient absorption of water and minerals. [2 marks]
  b Root hair cells can absorb mineral ions by diffusion and active transport.
    (i) Define the term *active transport*. [2 marks]
    (ii) Explain why respiration rates may increase in root hair cells during the uptake of mineral ions. [1 mark]

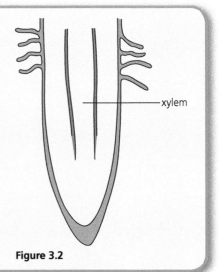

xylem

Figure 3.2

4 Biological molecules

Key objectives

The objectives for this chapter are to revise:

- the chemical elements that make up carbohydrates, fats and proteins
- the synthesis of large molecules from smaller base units
- food tests for starch, reducing sugars, fats, proteins and vitamin C

● Main nutrients

Nutrient	Elements present
Carbohydrate	Carbon, hydrogen, oxygen
Fat/oil (oils are liquid at room temperature but fats are solid)	Carbon, hydrogen, oxygen (but lower oxygen content than carbohydrates)
Protein	Carbon, hydrogen, oxygen, nitrogen, sometimes sulfur or phosphorus

Examiner's tips

- Carbohydrates, fats and proteins are all made up of the elements carbon, oxygen and hydrogen.
- Proteins also always contain nitrogen.
- One way of remembering the elements in carbohydrate is to look at its name: Carb *O* H*y*drate (carbon, oxygen, hydrogen).

Large carbohydrate molecules such as starch, glycogen and cellulose are made up of long chains of smaller units – monosaccharides such as glucose – held together by chemical bonds (Figure 4.1).

Fats are made up of three units of fatty acids chemically bonded to one glycerol unit (Figure 4.2).

Proteins are made of long chains of amino acids chemically bonded together (Figure 4.3). As there are about 20 different amino acids, their pattern in the chain can be quite complex, and the molecules can be very large.

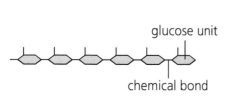

Figure 4.1 Carbohydrate

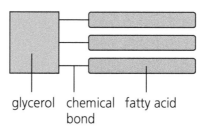

Figure 4.2 Fat

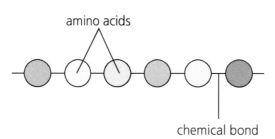

Figure 4.3 Protein

Cambridge IGCSE Biology Study and Revision Guide Second Edition © Dave Hayward 2016

Now try this

1 Make a table for carbohydrates, fats and proteins, with the headings shown below.   [9 marks]

Biological molecule	Chemical elements present	Sub-unit(s)	Examples

● Food tests

You need to be able to describe the tests for starch, reducing sugars, protein and fats. Make sure you also learn the colour change for a positive result.

Food tested	Name of test	Method	Positive result
Starch	Starch test (very original!)	Add a few drops of iodine solution to a solution of the food	Blue/black coloration
Reducing sugar	Benedict's test	Add an equal amount of Benedict's solution to a solution of the food. Boil carefully	A succession of colour changes: from turquoise to pale green, pea green, orange then brick red. The further the colour change is along the gradient, the more reducing sugar is present
Protein	Biuret test	Add an equal amount of sodium hydroxide to a solution of the food. Mix carefully. Then add a few drops of 1% copper sulfate, without shaking the mixture	Violet halo
Fats	Emulsion test	Dissolve the food in ethanol. Pour the solution into a clean test tube of water	White emulsion
Vitamin C	DCPIP test	DCPIP is a deep blue colour. Measure 2.0 cm^3 DCPIP solution into a test tube. Add fruit juice drop by drop, counting the drops used. Shake the mixture after each drop	DCPIP decolorises

● Common misconceptions

● Food tests do not all involve heating – the only food test that needs heating is the Benedict's test.

● Sample question

Describe how you could compare the vitamin C content of a range of fruits.
                                   [5 marks]

Student's answer

I would use the DCPIP test. ✓ First, I would get the juice from each of the fruits and measure out equal amounts. Add the juice to the DCIP drop by drop ✓ and record how many drops are needed to make the DCPIP change colour. Repeat for the other juices.

Examiner's comments

The answer could be improved by including more details, including the following:

- *The mixture should be shaken after each drop of DCPIP has been added.*
- *The colour change should be stated.*
- *A statement that the fewest drops needed indicates the juice containing the highest concentration of vitamin C.*
- *A statement about the need to clean out the glassware or pipette after each test.*

● Importance of water

Water is important to living things as a solvent – many substances (solutes) dissolve in it. Examples include glucose, salts and amino acids.

The different combinations and sequences of amino acids in a protein molecule can result in creating different three-dimensional shapes. Some proteins are dependent on their shape to perform a specific function. For example, the groove along an enzyme molecule is its active site (where it combines with a substrate molecule; see Chapter 5). If the shape of the enzyme molecule is changed, the substrate will no longer fit in the active site. Antibodies have a binding site with which they can attach themselves to an antigen with a particular shape (see Chapter 10).

● Structure of DNA

A DNA molecule is made up of two strands. Each strand is a chain of units called nucleotides. It can be thousands of nucleotides long.

Each nucleotide has one of four organic bases: A, T, C or G. Each base in one strand is cross-linked to a base in the other strand by a bond. A always bonds with T; C always bonds with G.

The double strand is twisted to form a double helix (a double spring).

● Importance of water as a solvent

- Most cells contain about 75% water.
- Many substances move around a cell when they are dissolved in water.
- Many important reactions take place in water.
- Food can be digested only when it is in liquid form – enzymes need to be in water to carry out their reaction.
- Excretory products such as urea are dissolved in water to dilute them and to allow their removal from the body.
- Transport systems, e.g. the bloodstream, are water based. Plasma is mainly water with substances dissolved in it (see Chapter 9).

(5) Enzymes

Key objectives

The objectives for this chapter are to revise:

- definitions of the key terms
- why enzymes are important in all living organisms
- enzyme action
- the effects of changes in temperature and pH on enzyme activity

- how to to explain enzyme action with reference to the active site, enzyme-substrate complex, substrate and product
- how to explain the specificity of enzymes
- how to explain the effect of changes in temperature and pH on enzyme activity

● Key terms

Catalyst	A substance that increases the rate of a chemical reaction and is not changed by the reaction
Enzyme	A protein that functions as a biological catalyst

● Enzymes and reactions

Most chemical reactions happening in living things are helped by enzymes. The speed at which they can catalyse reactions is sufficient to sustain life.

Most enzyme names end in -ase, e.g. lip*ase*, prote*ase*. Enzymes usually speed up reactions, but some slow them down. Some enzymes control reactions to build up molecules (synthesise them), e.g. starch phosphorylase:

$$\text{maltose} \xrightarrow{\text{starch phosphorylase}} \text{starch}$$

Others are involved in breaking down molecules, e.g. protease in digestion:

$$\text{protein} \xrightarrow{\text{protease}} \text{amino acids}$$

(See Chapter 7 for other examples of enzymes.)

Enzyme molecules are **proteins**. Each molecule has a special shape where the substrate fits (a **complementary shape**). Once the enzyme molecule and substrate come into contact, one or more products are formed.

Effect of temperature on enzymes

The optimum (best) temperature for enzyme-controlled reactions is around 37 °C (body temperature). Increasing the temperature above the optimum temperature slows the reaction down (see Figure 5.1).

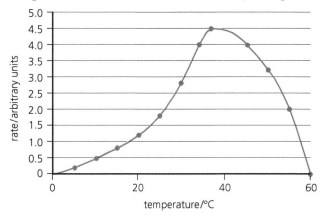

Figure 5.1 Graph showing the effect of temperature on the rate of an enzyme reaction

Effect of pH on enzymes

The pH of a solution is how acidic or alkaline it is. Most enzymes have an optimum pH (at which they work best) – usually around neutral (pH 7) – but there are some exceptions:

- pepsin, pH 2.0 – in the stomach, with hydrochloric acid;
- salivary amylase, pH 6.8 – in the mouth;
- catalase, pH 7.6 – in plants, e.g. potato;
- pancreatic lipase, pH 9.0 – in the duodenum.

The 'wrong' pH slows down enzyme activity, but this can usually be reversed if the optimum pH is restored.

Now try this

1 Make a table of the enzymes you have learned about. Use the headings shown below.   [12 marks]

Name of enzyme	Substrate (what the enzyme works on)	End product(s)	Other details (e.g. where reaction happens, optimum pH)

Examiner's tips

- You need to be able to state the definition of an enzyme – learn it by heart. Do this by reading the definition, then covering it up and writing it out. Then check it is correct. Repeat this three or four times. Test yourself again 24 hours later.
- Enzyme questions often involve plotting a set of results on a graph. Remember the key points for drawing a graph:
  - plot the independent variable (the figures you control) on the x- (horizontal) axis – these are the figures that usually go up in even stages;
  - label both axes with a title and units;
  - plot points in pencil (that way you can change any mistakes);
  - join the points with a line (this can be a curve).
- If you are instructed to predict a result using a graph, draw on the graph to read off the answer. Remember to state the units.

Enzymes are very specific (each chemical reaction is controlled by a different enzyme, so the enzyme has a high **specificity**). This is because of the shape of the **active site**. The shape of the active site of protease will be different from the shape of the active site of amylase, for example. This means that protease cannot break down starch and amylase cannot break down proteins. In other words, the enzyme is specific. For an enzyme to catalyse a reaction, the enzyme molecule and the **substrate** molecule need to meet and join together by means of a temporary bond. This temporary structure is called the **enzyme–substrate complex**. The **product** or products are then released and the enzyme molecule can combine with another substrate molecule.

Effect of temperature on enzymes

Enzymes work very slowly at low temperatures. This is because they have a low kinetic energy, so there are few collisions with the substrate molecules.

As the temperature is increased, the reaction rate increases because kinetic energy and, therefore, rate of collisions increases. However, above the optimum temperature for the enzyme, the reaction rate starts to decrease. This is because enzyme molecules start to permanently lose their shape at high temperature. This deforms the active site, so the enzyme and substrate cannot fit together (so no reaction). This effect is called denaturing. Most enzymes are **denatured** above 50 °C.

Effect of pH on enzymes

An enzyme works best at its optimum pH. At a higher or lower pH, the enzyme is less effective and an extreme pH can denature the enzyme – the active site is deformed permanently. This means there is no longer a complementary fit between the enzyme molecule and the substrate.

● Common misconceptions

- Enzymes are not denatured by low temperatures – they are just slowed down, and will work again when the temperature is suitable. Once an enzyme is denatured, the damage is permanent.

● Sample question

Figure 5.2 shows a box of biological washing powder. Study the information on the box.

1 Explain why:

  a the presence of protease and lipase would make the washing powder more effective than ordinary detergent   [3 marks]

  b the powder should not be used in boiling water.   [2 marks]

2 Silk is a material made from protein. Explain why the biological washing powder should not be used to wash silk clothes.   [2 marks]

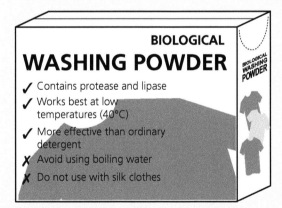

Figure 5.2

Student's answer

1 a Protease and lipase are enzymes, ✔ so they would break down stains ✔ better than ordinary detergent.
  b You could burn your hands when taking the clothes out. ✘
2 There is protease ✔ in the biological washing powder. This would digest the protein ✔ in the silk, so the clothes would get spoiled.

Cambridge IGCSE Biology Study and Revision Guide Second Edition © Dave Hayward 2016

Examiner's comments

1 **a** There are three marks available in this section. The candidate has made two valid statements, but has not given enough detail about what the enzymes digest (protease breaks down protein; lipase breaks down fats).

b The candidate does not answer the question – the statement needs to be related to the properties of enzymes – they are denatured at high temperatures.

2 This is a good response, gaining both the available marks.

Now try this

2 Six identical samples containing a mixture of starch and amylase in water were kept at different temperatures, and the time taken for the starch to be digested was measured. The results are shown in the following table.

Temperature at which samples were kept/°C	Time for starch to be digested/minutes
15	32
20	18
30	7
35	3
40	10
50	35

    a Plot a graph of these results.   [3 marks]
    b (i) Describe how the time taken for the starch to be digested could be determined.   [2 marks]
        (ii) At what temperature was the starch digested most rapidly?   [1 mark]
        (iii) Describe the relationship between temperature and the rate of starch digestion.   [2 marks]
    c Similar samples were set up and kept at 10 °C and 60 °C. The starch in these samples was not fully digested after one hour. Both of these samples were then kept at 35 °C. Suggest and explain the effect of the following changes in temperature on starch digestion:
        (i) sample changed from 10 °C to 35 °C   [2 marks]
        (ii) sample changed from 60 °C to 35 °C.   [2 marks]
3 a Adult female mosquitoes feed on the blood of mammals. They produce a protein-digesting enzyme called trypsin.
        (i) Explain why an adult female mosquito would need trypsin.   [2 marks]
        (ii) State the product that would be present in the gut of the mosquito if trypsin had been active.   [1 mark]
        (iii) State one use of this product in the body of an organism such as a mosquito.   [1 mark]
        Scientists have found a way of introducing a hormone into mosquitoes to switch off the trypsin secretion.
    b Suggest how this treatment would affect adult female mosquitoes.   [2 marks]
    c Enzymes such as trypsin are easily damaged. Suggest two ways of damaging an enzyme.   [2 marks]

6 Plant nutrition

Key terms

Photosynthesis	The process by which plants manufacture carbohydrates from raw materials using energy from light
Limiting factor	Something present in the environment in such short supply that it restricts life processes

Photosynthesis

The word equation for photosynthesis is:

$$\text{carbon dioxide + water} \xrightarrow[\text{chlorophyll}]{\text{light}} \text{glucose + oxygen}$$

Note: the glucose produced is converted to starch for storage in the leaf.

The equation shows that the raw materials for photosynthesis are carbon dioxide, water and light energy. The products are glucose (starch) and oxygen.

Common misconceptions

- Candidates often suggest that photosynthesis happens during the day and respiration happens at night. The first part of the sentence is correct, but the second part is wrong: respiration happens all the time (in both plants and animals).

The process of photosynthesis

1 Green plants take in carbon dioxide through their leaves. This happens by diffusion.

2 Water is absorbed through plants' roots by osmosis and transported to the leaf through xylem vessels.

3 Chloroplasts, containing chlorophyll, are responsible for trapping light energy. This energy is used to break up water molecules and then to bond hydrogen and carbon dioxide to form glucose.

4 Glucose is usually changed to sucrose for transport around the plant, or to starch for storage.

5 Oxygen is released as a waste product, or used by the plant for respiration.

Testing a leaf for starch

These tests show what factors are needed for photosynthesis and if it has taken place. Starch is stored in plant leaves as a product of photosynthesis. The starch test does not work by placing iodine solution on fresh leaves – it is not absorbed.

You need to be able to describe the starch test and the reasons for each stage. There are also some important safety points, outlined in the table below.

Stage	Reason	Safety points
Boil the leaf in water	To kill the leaf – this makes it permeable	Danger of scalding
Boil the leaf in ethanol	To decolorise the leaf – chlorophyll dissolves in ethanol	No naked flames – ethanol is highly flammable
Rinse the leaf in water	Boiling the leaf in ethanol makes it brittle – the water softens it	
Spread the leaf out on a white tile	So that the results are easy to see	
Add iodine solution to the leaf	To test for the presence of starch	Avoid skin contact with iodine solution

Factors needed for photosynthesis

Experiments can be used to find out what factors are needed for photosynthesis.

First, the plant is destarched. This involves leaving the plant in the dark for 48 hours. The plant uses up all the stores of starch in its leaves. One plant (or leaf) is exposed to all the conditions needed – this is the **control**. Another plant (or leaf) is deprived of one condition (this may be light or carbon dioxide).

After a few hours, the starch test is carried out on the control and test plant/leaf. The equation for photosynthesis shows the raw materials that a plant needs to make its food. Some plants have variegated leaves – only some parts of each leaf contain chlorophyll. When tested for starch, only the parts of the leaf with chlorophyll will contain starch.

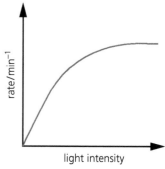

Figure 6.1

The carbon dioxide around a plant can be controlled by keeping the plant in a sealed container with a carbon dioxide absorber such as sodium hydroxide pellets. The control plant would be in an identical container, without the carbon dioxide absorber.

When light intensity, carbon dioxide or temperature is increased, the rate of photosynthesis increases. However, there comes a point when further increases do not increase the rate (see Figure 6.1).

Now try this

1 Figure 6.2 shows a variegated leaf in a photosynthesis experiment. Part of the leaf has been covered with black paper. The leaf was then exposed to light for a few hours. Leaf discs were then cut from regions of the leaf at A, B, C and D. Each disc was tested for the presence of starch.
  Predict the results of the starch test on each region of the variegated leaf. Give a reason for each result. [8 marks]

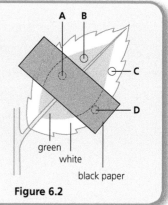

Figure 6.2

Cambridge IGCSE Biology Study and Revision Guide Second Edition © Dave Hayward 2016

It is difficult to show that land plants produce oxygen during photosynthesis because the gas diffuses into the air. However, some aquatic plants produce bubbles of oxygen. These can be collected and tested with a glowing splint – this relights in oxygen.

For the Extended paper, you need to learn the balanced chemical equation for photosynthesis.

Chemical equation:

$$6CO_2 + 6H_2O \xrightarrow[\text{chlorophyll}]{\text{light}} C_6H_{12}O_6 + 6O_2$$

Chlorophyll traps light energy and transfers it into chemical energy in molecules for the synthesis of carbohydrates. First, glucose is formed ($C_6H_{12}O_6$ in the equation). However, this is converted into sucrose for translocation around the plant. The sucrose is changed to starch for storage. This is insoluble and causes no osmotic problems. Other carbohydrates, e.g. cellulose for making cell walls, can also be synthesised from sucrose. Sucrose is also an energy source for the plant.

● Sample question

The chemical equation for photosynthesis shown below is incomplete.

$$6H_2O + \underline{\hspace{3cm}} \xrightarrow[\text{plant pigment}]{\text{energy}} C_6H_{12}O_6 + \underline{\hspace{3cm}}$$

1 Complete the equation in either all words or all symbols. [2 marks]

2 State the source of energy for this reaction. [1 mark]

3 Name the plant pigment necessary for this reaction. [1 mark]

Student's answer

1 $6H_2O + 6CO_2$ ✔ $\xrightarrow[\text{plant pigment}]{\text{energy}} C_2H_2O_2 + $ oxygen ✖

2 The Sun ✖
3 Chloroplast ✖

Examiner's comments

1 Although the candidate has identified both compounds missing from the equation, she has written one in symbols and the second in words, so the examiner has not awarded the second mark. When you write an equation or complete one, always do it with either words or symbols – do not mix them up.

2 'The Sun' has not been accepted because this answer is not specific enough – the Sun produces two types of energy (light and heat). Plants use only light energy in photosynthesis. The correct answer was light or sunlight.

3 Chloroplasts are structures, not pigments. They contain the pigment. Chlorophyll was the correct answer.

Factors affecting the rate of photosynthesis

As light intensity increases, so does the rate of photosynthesis. This can be demonstrated as shown in Figure 6.1 using an aquatic plant such as *Elodea*.

The light intensity (I) is related to the distance (d) between the lamp and the plant ($I = 1 / d^2$). As the lamp is moved closer, the light intensity increases. The rate of photosynthesis is directly proportional to the light intensity, as shown by the graph in Figure 6.1. However, the photosynthetic rate cannot be increased indefinitely: a point is reached where all the chloroplasts cannot trap any more light.

In addition, if there is a limiting factor (such as carbon dioxide), the rate of photosynthesis becomes limited, as shown in Figure 6.3.

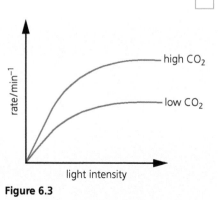

Figure 6.3

Glasshouse systems

Glasshouses are used in some countries to control conditions for plant growth, especially when growing conditions outside are not ideal. The glass helps trap heat inside and atmospheric conditions can be controlled.

Carbon dioxide enrichment

Atmospheric air contains only 0.04% carbon dioxide, so it can easily become a factor that limits the rate of photosynthesis. A glasshouse is a closed system, so the content of the air in it can be controlled. For example, the amount of carbon dioxide can be increased by burning fossil fuels in the greenhouse, or releasing pure carbon dioxide from a gas cylinder.

Optimum light

If light conditions in a glasshouse are not optimum (e.g. in winter), they can be improved by using artificial lights.

Optimum temperature

If the temperature is a limiting factor, e.g. in winter, it can be raised by using a heating system. If fossil fuels are burned, there is also a benefit from the carbon dioxide produced.

Hydrogencarbonate indicator can be used to investigate the effect of gas exchange of an aquatic plant kept in the light and in the dark. Fresh indicator is a pink/red colour. If there is a build-up of carbon dioxide (no photosynthesis, but respiration is producing carbon dioxide), the decrease in pH turns the indicator yellow. If the carbon dioxide level drops (e.g. during photosynthesis), the indicator turns purple.

> ## Now try this
>
> 2 Figure 6.4 shows the rates of sugar production by a plant on a bright day and on a dull day.
>   a (i)   Which curve, A or B, shows sugar production on a bright day? State a reason for your choice. [2 marks]
>   (ii) Outline the role of chloroplasts in photosynthesis. [2 marks]
>   (iii) Suggest one feature of a leaf, other than the presence of chloroplasts, that might affect the amount of sugar produced. [1 mark]

b (i) Suggest why many plants store starch rather than sugars. [1 mark]
  (ii) Which chemical reagent is used to test for starch? [1 mark]

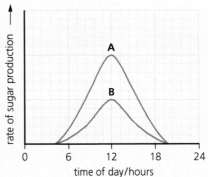

Figure 6.4

3 A student carried out an experiment to investigate the growth of floating water plants taken from a pond. Equal masses of the plants were placed into three separate glass containers – A, B and C – similar to that shown in Figure 6.5. Container A was lit by a 250-W bulb, container B was lit by a 75-W bulb and container C was lit by a 250-W bulb with a coloured filter in front of the lamp, as shown in Figure 6.5. At weekly intervals, the plants were removed from each container in turn, gently dried, weighed and returned to the containers after their mass had been recorded. Figure 6.6 shows the results.

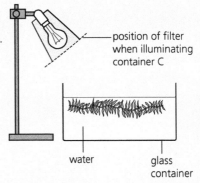

Figure 6.5

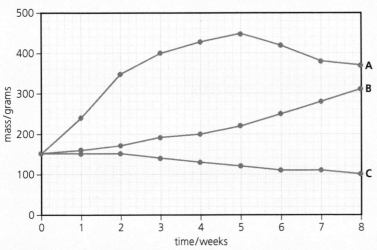

Figure 6.6

a Calculate the percentage increase in mass of the plants in container A during the first five weeks of the experiment. (Show your working.) [2 marks]

b Suggest why the mass of plants in container A began to decrease after week 5, while the mass of plants in container B continued to increase. [2 marks]

c During the eighth week, in which container would there be the least dissolved oxygen? Explain your answer. [2 marks]

Figure 6.7 shows the amount of light of different colours absorbed by chlorophyll. The filter used in illuminating container C allowed only one colour of light to pass through to the water plants.

d Suggest which colour of light passed through the filter. Explain your answer. [2 marks]

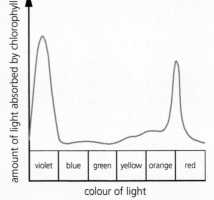

Figure 6.7

Leaf structure

You need to be able to identify the cellular and tissue structure of a leaf of a dicotyledonous plant.

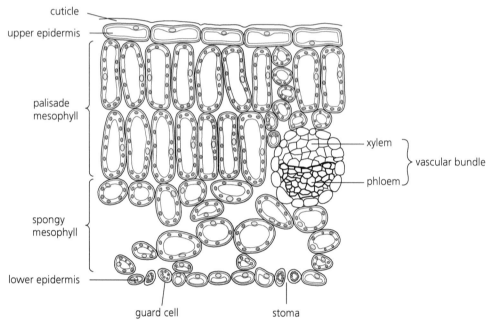

Figure 6.8 A cross-section through part of a leaf

Now try this

4 Copy, e.g. by tracing, the diagram of a leaf in Figure 6.8. *Practise* labelling the structures present. Annotate the diagram by writing one statement about each of the labels.   [7 marks]

The table below explains how each feature of the internal structure of a leaf is adapted for photosynthesis.

Part of leaf	Details
Cuticle	Made of wax, waterproofing the leaf. It is secreted by cells of the upper epidermis
Upper epidermis	Thin and transparent cells that allow light to pass through. No chloroplasts are present. They act as a barrier to disease organisms
Palisade mesophyll	Main region for photosynthesis. Cells are columnar (quite long) and packed with chloroplasts to trap light energy. They receive carbon dioxide by diffusion from air spaces in the spongy mesophyll
Spongy mesophyll	Cells are more spherical and loosely packed. They contain chloroplasts, but not as many as in palisade cells. Air spaces between cells allow gaseous exchange – carbon dioxide to the cells, oxygen from the cells – during photosynthesis
Vascular bundle	A leaf vein made up of xylem and phloem. Xylem vessels bring water and minerals to the leaf. Phloem vessels transport sugars and amino acids away (this is called translocation)
Lower epidermis	Acts as a protective layer. Stomata are present to regulate the loss of water vapour (this is called transpiration). Site of gaseous exchange into and out of the leaf
Stomata	Each stoma is surrounded by a pair of guard cells. These can control whether the stoma is open or closed. Water vapour passes out during transpiration. Carbon dioxide diffuses in and oxygen diffuses out during photosynthesis

Adaptations of a leaf for photosynthesis

- Their broad, flat shape offers a large surface area for absorption of sunlight and carbon dioxide.
- Most leaves are thin and the carbon dioxide has to diffuse across only short distances to reach the inner cells.
- The large spaces between cells inside the leaf provide an easy passage through which carbon dioxide can diffuse.
- There are many stomata (pores) in the lower surface of the leaf. These allow the exchange of carbon dioxide and oxygen with the air outside.

Cambridge IGCSE Biology Study and Revision Guide Second Edition   © Dave Hayward 2016

- There are more chloroplasts in the upper (palisade) cells than in the lower (spongy mesophyll) cells. The palisade cells, being on the upper surface, will receive most sunlight and this will reach the chloroplasts without being absorbed by too many cell walls.
- The branching network of veins provides a good water supply to the photosynthesising cells.

Nitrate and magnesium ions

Nitrate ions are needed for making amino acids. These are the building blocks of proteins. Remember that all proteins contain the element nitrogen (see Chapter 4). Each amino acid is formed by combining sugars, made during photosynthesis, with nitrate. The amino acids are made into long chains by bonding them together. The proteins are used to make cytoplasm and enzymes.

Magnesium ions are needed to make chlorophyll. Each chlorophyll molecule contains one magnesium atom. Plants need chlorophyll to trap light to provide energy during photosynthesis.

Nitrate ion and magnesium ion deficiency

You already know the importance of nitrate ions for protein synthesis. If the plant has a nitrate ion deficiency, it will not be able to make proteins, so growth will slow down. The stem becomes weak and lower leaves become yellow and die, while upper leaves turn pale green.

You already know the importance of magnesium ions for synthesis of chlorophyll. If the plant has a magnesium ion deficiency, it will not be able to make chlorophyll. Leaves turn yellow from the bottom of the stem upwards. Plant growth will suffer because it will have reduced photosynthesis. Yellowing of leaves due to lack of magnesium ions is called chlorosis.

Cambridge IGCSE Biology Study and Revision Guide Second Edition © Dave Hayward 2016

(7) Human nutrition

Key objectives

The objectives for this chapter are to revise:

- definitions of the key terms
- how age, gender and activity affect the dietary needs of humans
- the effects of malnutrition
- the main sources and the importance of the main foodstuffs
- the treatment of diarrhea
- how to describe cholera
- the main regions of the alimentary canal and the functions of its main organs
- the types and structure of human teeth, their functions, the causes of dental decay and the proper care of teeth

- the significance of chemical digestion in the alimentary canal
- the sites of secretion and functions of key enzymes, listing substrates and end products
- the functions of hydrochloric acid in gastric juice
- the absorption of digested food and water
- the causes and effects of vitamin D and iron deficiencies and protein-energy malnutrition
- the effects of the cholera bacterium
- the digestion of starch in the alimentary canal
- pepsin and trypsin as proteases and their sites of action
- the role of bile in neutralisation and emulsification
- the significance of villi in the small intestine, their structure and the role of capillaries and lacteals

● Key terms

Balanced diet	This is a diet that contains all the essential nutrients in the correct proportions to maintain good health. The nutrients needed are carbohydrate, fat, protein, vitamins, minerals, fibre and water
Ingestion	The taking of substances such as food and drink into the body through the mouth
Mechanical digestion	The breakdown of food into smaller pieces without chemical change to the food molecules
Chemical digestion	The breakdown of large insoluble molecules into small soluble molecules
Absorption	The movement of small food molecules and ions through the wall of the intestine into the blood
Assimilation	The movement of digested food molecules into the cells of the body where they are used, becoming part of the cells
Egestion	The passing out of food that has not been digested or absorbed, as faeces, through the anus

● Dietary requirements

Your dietary requirements depend on your age, sex and levels of physical activity. The amount of energy needed is provided mainly by our carbohydrate and fat intake. Generally, males use up more energy than females, and energy demand increases until we stop growing. Someone doing physical work will use up more energy than an office worker. While children are growing, they need more protein per kilogram of body weight than adults do. Pregnant women need extra nutrients for the development of the fetus. Once the baby has been born, a breast-feeding mother will need more protein and minerals, e.g. calcium, in her diet to satisfy the baby's requirements.

● Effects of malnutrition

Malnutrition is the result of an unbalanced diet:

- Too much food – or too much carbohydrate, fat or protein – can lead to obesity. This can lead to coronary heart disease and diabetes (which can cause blindness).

Cambridge IGCSE Biology Study and Revision Guide Second Edition © Dave Hayward 2016

- Too much animal fat in the diet results in high cholesterol levels. Cholesterol can stick to the walls of arteries, gradually blocking them. If coronary arteries become blocked, the result can be angina and coronary heart disease.
- Too little food can result in starvation. Extreme slimming diets, such as those that avoid carbohydrate foods, can result in the disease anorexia nervosa.
- Constipation is caused by a lack of fibre in the diet. It can lead to bowel cancer.
- Vitamin and mineral deficiency diseases are all the result of malnutrition. Scurvy is caused by a lack of vitamin C.

● Main nutrients

Nutrient	Use in the body	Good food sources
Carbohydrate	Source of energy	Rice, potato, sweet potato, cassava, bread, millet, sugary foods (e.g. cake, jam, honey)
Fat/oil (oils are liquid at room temperature but fats are solid)	Source of energy (twice as much as carbohydrate); used as insulation against heat loss, for some hormones, in cell membranes, for insulation of nerve fibres	Butter, milk, cheese, egg yolk, animal fat, groundnuts (peanuts)
Protein	Growth, tissue repair, enzymes, some hormones, cell membranes, hair, nails. Can be broken down to provide energy	Meat, fish, eggs, soya, groundnuts, milk, meat substitute (e.g. Quorn), cowpeas
Vitamin C	Needed to maintain healthy skin and gums	Citrus fruits, blackcurrants, cabbage, tomato, guava, mango
Vitamin D	Needed to maintain hard bones. Helps in absorption of calcium from small intestine	Milk, cheese, egg yolk, fish liver oil. Can be made in the skin when exposed to sunlight
Calcium	Needed to form healthy bones and teeth and for normal blood clotting	Milk, cheese, fish
Iron	Needed for formation of haemoglobin in red blood cells	Red meat, liver, kidney, eggs, green vegetables (spinach, cabbage, cocoyam, groundnut leaves), chocolate
Fibre	This is cellulose. It adds bulk to undigested food passing through the intestines, maintaining peristalsis	Vegetables, fruit, wholemeal bread
Water	Formation of blood, cytoplasm, as a solvent for transport of nutrients and removal of wastes (as urine). Enzymes work only in solution	Drinks, fruit, vegetables

Vitamins and minerals, although needed in only small quantities, are important for maintaining good health. A shortage can result in a deficiency disease. You only need to know vitamins C and D, and the minerals calcium and iron. Fibre (roughage) is needed in much larger quantities. Do not forget that water is also a vital part of our dietary requirements.

Now try this

1 Make a mnemonic to remember all the nutrients in a balanced diet. You need to use the letters C, F, P, V, M, F, W. [2 marks]

Cambridge IGCSE Biology Study and Revision Guide Second Edition © Dave Hayward 2016

● Sample question

The following table shows the carbohydrate content of some vegetables.

Vegetable	Total carbohydrate, g/100 g	Starch, g/100 g	Fibre, g/100 g
Beans	15.1	9.3	3.5
Broccoli	1.1	Trace	2.3
Cabbage	4.1	0.1	2.4
Carrots (boiled)	4.9	0.2	2.5
Chickpeas	18.2	16.6	4.3
Onions	3.7	Trace	0.7
Peas (frozen, boiled)	9.7	4.7	5.1
Potato (boiled)	17.0	16.3	1.2
Sweet potato (boiled)	20.5	8.9	2.3
Tomatoes (raw)	3.1	Trace	1.0

1 Name the chemical elements present in a carbohydrate.

2 State which vegetable in the table contains:

   a the highest proportion of total carbohydrate

   b the highest proportion of fibre.

3 Total carbohydrate is calculated as the sum of starch and sugars in the vegetable.

   a Name the vegetable that contains the highest proportion of sugar per 100 g vegetable.

   b Calculate the amount of sugar present in 500 g of the vegetable named in 3a. Show your working.

Student's answer

1 C, H, O ✖ [1 mark]
2 a Sweet potato ✔ [1 mark]
  b Peas ✔ [1 mark]
3 a Sweet potato ✔ [1 mark]
  b 20.5 − 8.9 = 11.6
    11.6 × 5 ✔ = 58 ✖ [2 marks]

Examiner's comments

Most of the answers were good, but this candidate made two easily avoidable errors:

1 Do not use abbreviations such as symbols when you are asked to name elements.

3 b Remember to state the units when giving the answer to a calculation. This candidate gained one mark for showing the correct working for the calculation, but lost the second mark because of the lack of units – which should have been 'g' (grams).

Now try this

2 The chart in Figure 7.1 is used to find a person's ideal mass.

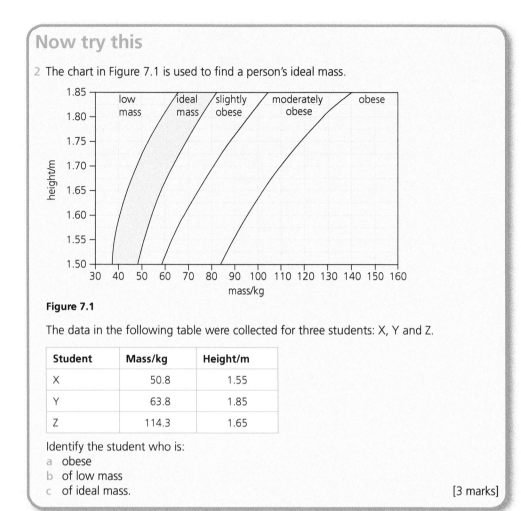

Figure 7.1

The data in the following table were collected for three students: X, Y and Z.

Student	Mass/kg	Height/m
X	50.8	1.55
Y	63.8	1.85
Z	114.3	1.65

Identify the student who is:
a obese
b of low mass
c of ideal mass.   [3 marks]

● Vitamin D and iron deficiencies

A shortage of **vitamin D** can lead to a deficiency disease called **rickets**. The symptoms are soft bones that become deformed. Sufferers may become bow legged. Exposure to moderate sunlight helps the body make vitamin D. Thus, a lack of exposure (because of climate or season or wearing clothing that acts as a barrier to sunlight) can result in the development of a deficiency.

A deficiency of **iron** in the diet can lead to **anaemia**. The symptoms are constant tiredness and a lack of energy. Normally, as red blood cells in the body are broken down, the iron in the haemoglobin is recycled to make new red blood cells. However, if a woman has a heavy period (see Chapter 16) there is a lot of blood (and therefore iron) loss, which can result in anaemia.

● Protein–energy malnutrition

Sometimes the balance of food in the diet is wrong, e.g. too much carbohydrate and too little protein, such as when the bulk of the diet is starchy food, such as sweet potato or cassava. This can lead to **kwashiorkor** in young children. They lack protein, but other problems such as plant toxins can also play a role. The symptoms of kwashiorkor are dry skin, pot-belly, changes to hair colour, weakness and irritability.

Marasmus is an acute form of malnutrition. The condition is caused by a very poor diet with inadequate carbohydrate intake and a lack of protein. The symptoms are emaciation, with reduced fat and muscle tissue. The skin is thin and hangs in folds.

● Alimentary canal

Figure 7.2 shows the main organs of the alimentary canal. The table below gives their functions.

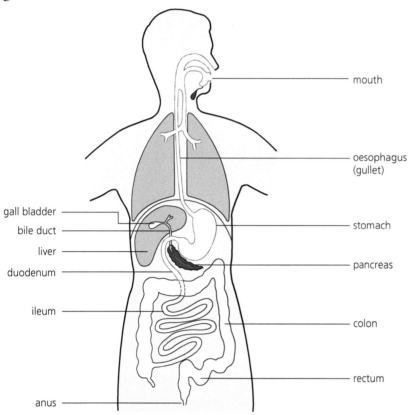

Figure 7.2

Region	Function
Mouth	Food is ingested here. It is mechanically digested by cutting, chewing and grinding of teeth. Saliva is added
Salivary glands	Produce saliva, containing the enzyme amylase to begin the chemical digestion of starch. The water in saliva helps lubricate food and makes small pieces stick together
Oesophagus	Boluses (balls) of food pass through by peristalsis, from mouth to stomach
Stomach	Muscular walls squeeze on food to make it semi-liquid. Gastric juice contains protease to chemically digest protein and hydrochloric acid to maintain an optimum pH (1–2.5). The acid also kills bacteria
Duodenum	This is the first part of the small intestine. It receives pancreatic juice containing protease, lipase and amylase. The juice also contains sodium hydrogencarbonate, which neutralises acid from the stomach, producing a pH of 7–8
Ileum	The second part of the small intestine. Enzymes in the epithelial lining chemically digest maltose and peptides. Its surface area is increased by the presence of villi, which allow the efficient absorption of digested food molecules
Pancreas	Secretes pancreatic juice into the duodenum for chemical digestion of proteins, fats and starch
Liver	Makes bile, which is stored in the gall bladder. Bile contains salts that emulsify fats, forming droplets with a large surface area to make digestion by lipase more efficient. Digested foods are assimilated here. For example, glucose is stored as glycogen; surplus amino acids are deaminated (see Chapter 13)
Gall bladder	Stores bile, made in the liver, to be secreted into the duodenum via the bile duct
Colon	The second part of the large intestine. Its main function is the reabsorption of water from undigested food. It also absorbs bile salts to pass back to the liver
Rectum	Stores faeces until they are egested
Anus	Has muscles to control when faeces are egested from the body

● Common misconceptions

● The liver does not make digestive enzymes – bile is not an enzyme. It breaks fat down into smaller droplets, but does not change them chemically. The fat molecules stay the same size; it is just the droplet size that changes from large to small due to the action of bile.

● Diarrhoea

Diarrhoea is the loss of watery faeces. It is sometimes caused by bacterial or viral infection, for example from food or water, resulting in the intestines being unable to absorb fluid from the contents of the colon or too much fluid being secreted into the colon. Undigested food then moves through the large intestine too quickly, resulting in insufficient time to absorb water from it. Unless the condition is treated, **dehydration** can occur.

The treatment is called **oral hydration therapy** – drinking plenty of fluids (sipping small amounts of water at a time) to rehydrate the body.

● Cholera

This disease is caused by a bacterium that causes acute diarrhoea.

Effects of the cholera bacteria

When cholera bacteria, *Vibrio cholerae*, are ingested they multiply in the small intestine and invade its epithelial cells. As the bacteria become embedded, they release toxins (poisons) that irritate the intestinal lining and lead to the secretion of large amounts of water and salts, including chloride ions. The salts decrease the osmotic potential of the gut contents, drawing more water from surrounding tissues and blood by osmosis (see 'Osmosis' in Chapter 3). This makes the undigested food much more watery, leading to acute diarrhoea, and the loss of body fluids and salt leads to dehydration and kidney failure.

● Mechanical digestion

Teeth are involved in the mechanical digestion of food. You need to know about the types and functions of human teeth and also about tooth structure. Figure 7.3 shows the types and functions of human teeth.

	Incisor	Canine	Premolar	Molar
	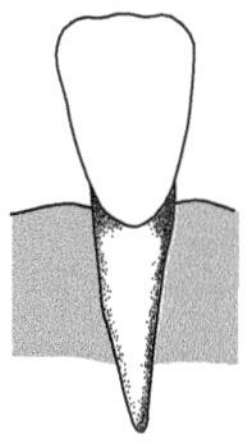	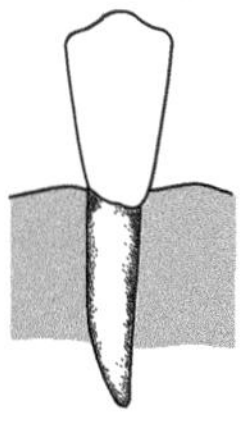	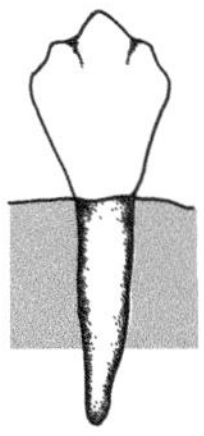	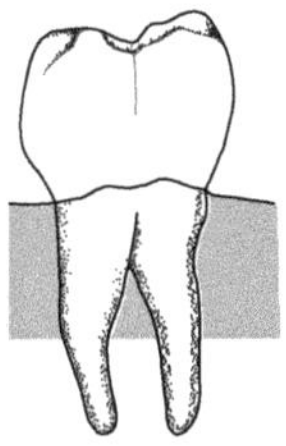
Position in mouth	Front	Either side of incisors	Behind canines	Back
Description	Chisel shaped (sharp edge)	Slightly more pointed than incisors	Have two points (cusps). Have one/two roots	Have four/five cusps. Have two/three roots
Function	Biting off pieces of food	Similar function to incisors	Tearing and grinding food	Chewing and grinding food

Figure 7.3

Cambridge IGCSE Biology Study and Revision Guide Second Edition © Dave Hayward 2016

Structure of a tooth

Figure 7.4 shows a section through a molar tooth.

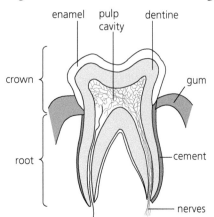

Figure 7.4

Causes of dental decay

Bacteria are present on the surface of our teeth. Food deposits and bacteria form a layer called **plaque**. Bacteria in the plaque respire sugars, producing acid. This acid dissolves enamel and dentine, forming a hole. Dentine underneath the enamel is softer – it dissolves more rapidly. If the hole reaches the pulp cavity, bacterial infection can get to the nerve. This results in toothache and, possibly, an abscess (an infection in the jaw).

● Common misconceptions

● Do not say that sugar causes tooth decay. It only causes problems because of the activity of bacteria feeding on it and producing acids.

Now try this

3 Figure 7.5 shows the four types of teeth found in humans.
  a Copy the figure and label one example of each of the four types of teeth.   [4 marks]
  b (i) What is the function of the teeth labelled A?   [1 mark]
      (ii) What is the function of the teeth labelled B?   [1 mark]
  c The outer layer of the crown of a tooth is resistant to attack by bacteria.
    (i) Name this outer layer.   [1 mark]

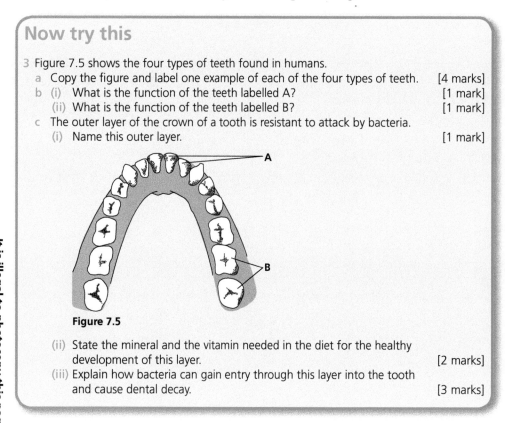

Figure 7.5

    (ii) State the mineral and the vitamin needed in the diet for the healthy development of this layer.   [2 marks]
    (iii) Explain how bacteria can gain entry through this layer into the tooth and cause dental decay.   [3 marks]

Cambridge IGCSE Biology Study and Revision Guide Second Edition   © Dave Hayward 2016

Proper care of teeth

- Avoid sugary foods, especially between meals, so bacteria cannot make acid; in addition, clean teeth regularly to remove plaque.
- Use dental floss or a toothpick to remove pieces of food and plaque trapped between teeth.
- Use a fluoride toothpaste (or drink fluoridated water) – fluoride hardens tooth enamel.
- Visit a dentist regularly to make sure any tooth decay is treated early and any stubborn plaque (called calculus) is removed.

● Chemical digestion

Food that we ingest is mainly made up of large, insoluble molecules that cannot be absorbed through the gut wall. It needs to be changed into small, soluble molecules.

Chemical digestion involves breaking down large, insoluble molecules into small, soluble ones. Enzymes speed up the process. They work efficiently at body temperature (37 °C) and at a suitable pH. The main places where chemical digestion happens are the mouth, stomach and small intestine.

You need to be able to state the functions of amylase, protease and lipase.

Enzyme	Site of action	Special conditions	Substrate digested	End product(s)
Amylase	Mouth, duodenum	Slightly alkaline	Starch	Maltose, glucose
Protease	Stomach, duodenum	Acid in stomach, alkaline in duodenum	Protein	Amino acids
Lipase	Duodenum	Alkaline	Fat	Fatty acids and glycerol

You need to be aware of the functions of hydrochloric acid in gastric juice (in the stomach):

- It gives an acid pH, which protease needs to work at its optimum.
- It kills bacteria that may be present in food that has been ingested.

● Common misconceptions

- Chewing food does not involve breaking down large molecules into small molecules; it only breaks down large pieces into smaller pieces, giving a larger surface area for enzymes to work on.

Digestion of starch

Starch is digested in two places in the alimentary canal: by salivary amylase in the mouth and by pancreatic amylase in the duodenum. Amylase works best in a neutral or slightly alkaline pH and converts large, insoluble starch molecules into smaller, soluble maltose molecules. Maltose is a disaccharide sugar and is still too big to be absorbed through the wall of the intestine. Maltose is broken down to glucose by the enzyme **maltase**, which is present in the membranes of the epithelial cells of the villi.

Digestion of protein

There are actually several proteases that break down proteins. One protease is **pepsin**, which is secreted in the stomach. Pepsin acts on proteins and breaks them down into soluble compounds called peptides. These are shorter chains of amino acids than proteins. Another protease is called **trypsin**. Trypsin is secreted by the pancreas in an inactive form, which is changed to an active enzyme in the duodenum (part of the small intestine). It has a similar role to pepsin, breaking down proteins to peptides.

Hydrochloric acid in gastric juice

The hydrochloric acid that is secreted by cells in the wall of the stomach creates a very acid pH of 2. This pH is important because it denatures enzymes in harmful organisms in food, such as bacteria (which may otherwise cause food poisoning), and it provides the optimum pH for the protein-digesting enzyme pepsin to work.

Bile

Bile is made in the liver, stored in the gall bladder and transferred to the duodenum by the bile duct (Figure 7.2). It has no enzymes but does contain bile salts, which act on fats in a similar way to a detergent.

The bile salts **emulsify** the fats, breaking them up into small droplets with a large surface area, which are more efficiently digested by lipase.

Bile is slightly alkaline, as it contains sodium hydrogencarbonate, and has the function of neutralising the acidic mixture of food and gastric juices as it enters the duodenum. This is important because enzymes secreted into the duodenum need alkaline conditions to work at their optimum rate.

● Absorption

The **small intestine** has a very rich blood supply. Digested food molecules are small enough to pass through the wall of the intestine into the bloodstream.

The small intestine and the colon are both involved in the absorption of water, but the small intestine absorbs the most.

Role of villi in absorption

You need to be able to relate the structure of the small intestine to its function of absorbing digested food and describe the significance of villi in increasing the internal surface area.

Villi are present in the small intestine – these are finger-like projections that increase the surface area for absorption. If a section of small intestine was turned inside out, its surface would be like a carpet. The surface area of a villus is further increased by the presence of microvilli.

Inside each **villus** are blood capillaries that absorb amino acids and glucose. There are also **lacteals** – these absorb fatty acids and glycerol.

Food molecules are absorbed mainly by diffusion. Figure 7.6 shows the features of a villus that increase the efficiency of diffusion. Molecules can also be absorbed by active transport (see Chapter 3). Epithelial cells contain mitochondria to provide energy for absorption against the concentration gradient.

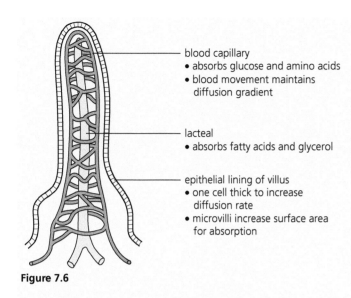

blood capillary
• absorbs glucose and amino acids
• blood movement maintains diffusion gradient

lacteal
• absorbs fatty acids and glycerol

epithelial lining of villus
• one cell thick to increase diffusion rate
• microvilli increase surface area for absorption

Examiner's tips

Absorption in the small intestine involves diffusion and active transport (described in Chapter 3). Check that you can relate how the structure of a villus makes these processes efficient.

Figure 7.6

● Sample question

1 Proteins are digested in the stomach and small intestine.

   a Which type of enzyme breaks down proteins?   [1 mark]

   b State how the conditions necessary for the digestion of proteins in the stomach are different from those in the small intestine.   [1 mark]

2 When carbohydrates have been digested, excess glucose is stored.

   a Where is it stored?   [1 mark]

   b What is it stored as?   [1 mark]

3 Excess amino acids cannot be stored. Describe how they are removed from the body.   [4 marks]

Student's answer

1 a Protease ✔
   b It is acid in the stomach and alkaline in the small intestine. ✔
2 a In the liver ✔
   b Glucagon ✖
3 The liver ✔ breaks them down. This makes urea. ✔ The kidney filters out the urea. ✔

Examiner's comments

Answers to parts 1 and 2 are good, except that the candidate has got glucagon (a hormone) mixed up with glycogen (the correct answer). Names such as this have to be accurately spelt. The answer for part 3 contains only three statements. Further details about the filtering of blood or the formation of urine and its removal by urination would have achieved the final mark.

Cambridge IGCSE Biology Study and Revision Guide Second Edition © Dave Hayward 2016

8 Transport in plants

● Key terms

Transpiration	The loss of water vapour from plant leaves by evaporation of water at the surfaces of the mesophyll cells followed by the diffusion of water vapour through the stomata
Translocation	Movement of sucrose and amino acids in the phloem from regions of production (source) to regions of storage or to regions where they are used for respiration or growth (sink)

● Transport in plants

Two types of tissues are present in plants to transport materials. The **xylem** carries water and salts, as well as providing support for the plant. The **phloem** carries food substances – sugars and amino acids.

Xylem and phloem are found in vascular bundles in roots (Figure 8.1), stems (Figure 8.2) and leaves (Figure 6.8).

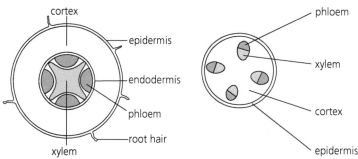

Figure 8.1 Section through a root

Figure 8.2 Section through a stem

● Water uptake

Root hair cells form on young roots to increase the surface area of the root for absorption of water and mineral ions, as well as providing anchorage for the plant. Figure 8.3 shows a root hair cell.

The cell extension (the hair) increases the surface area of the cell to make it more efficient in absorbing materials.

Figure 8.3

Cambridge IGCSE Biology Study and Revision Guide Second Edition © Dave Hayward 2016

1 a Copy or trace the root hair cell in Figure 8.3. Label all its parts, remembering to use a
   ruler for all label lines.   [5 marks]
  b Which plant cell part is missing from this cell?   [1 mark]
  c Name the process by which the cell absorbs:
    (i) water   [1 mark]
    (ii) mineral ions.   [1 mark]

Passage of water through root, stem and leaf

Water passes through the root hair cells to the root cortex cells by osmosis, reaching the xylem vessels in the centre (see Figure 8.1).

When water reaches the xylem, it travels up these vessels through the stem to the leaves. Mature xylem cells have no cell contents, so they act like open-ended tubes allowing free movement of water through them. In the leaves, water passes out of the xylem vessels into the surrounding cells (mesophyll cells). Mineral ions are also transported through the xylem.

Root hair cells have a large surface area and they are very numerous on young roots. This increases the rate of the absorption of water by osmosis and ions by active transport.

Transpiration

Transpiration is the loss of water vapour from a leaf. Water in the leaf mesophyll cells forms a thin layer on their surfaces. The water evaporates into the air spaces in the spongy mesophyll. This creates a high concentration of water molecules. They diffuse out of the leaf into the surrounding air, through the stomata, by diffusion.

Factors affecting transpiration rate

The following table shows two factors that can affect the rate of transpiration. If these factors are reversed, the rate will also reverse.

Factor	Effect on transpiration rate
Increase in temperature	Increases transpiration rate
Increase in humidity	Decreases transpiration rate

Examiner's tip

Most of the factors that result in a change in transpiration rate are linked to diffusion. When writing explanations, try to include references to the concentration gradient caused by a change in the factor.

2 Figure 8.4 shows part of the lower surface of a typical dicotyledonous leaf.

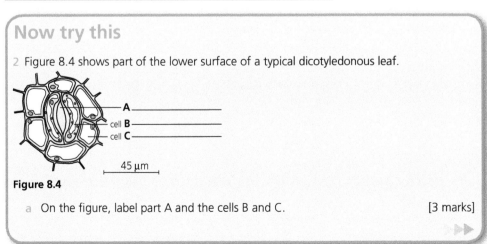

45 µm

Figure 8.4

  a On the figure, label part A and the cells B and C.   [3 marks]

▶▶

The surfaces of the leaves of two species of plant were studied and the number of stomata per unit area (stomatal frequency) was recorded.
Cobalt chloride paper changes colour in the presence of water.
Pieces of cobalt chloride paper were attached to the upper and lower surfaces of leaves on both plants. The plants were set up for one hour during the day. Any colour changes were recorded. The experiment was repeated for one hour at night. The table shows the results.

Plant species	Stomatal frequency		Colour change to cobalt chloride paper			
	Lower surface	Upper surface	Day		Night	
			Lower surface	Upper surface	Lower surface	Upper surface
Cassia fistula	0	18	✖	✔	✖	✖
Bauhinia monandra	22	0	✔	✖	✖	✖

Key: ✔, colour change; ✖, no colour change.

b Describe the differences in stomatal distribution between the two species of plant. [2 marks]
c (i) Explain the colour changes to the cobalt chloride paper during the day. [3 marks]
   (ii) Suggest why there was no colour change for either plant at night. [1 mark]
d Outline the mechanism by which water in the roots reaches the leaf. [3 marks]
e State and explain the effect of the following on transpiration rate:
   (i) increasing humidity [2 marks]
   (ii) increasing temperature. [2 marks]

The loss of water vapour in the process of transpiration is related to the large surface area represented by the cell surfaces in the mesophyll of the leaf, the interconnecting air spaces in the spongy mesophyll and the stomata.

Mechanism of water uptake

Water enters root hair cells by osmosis. This happens when the water potential in the soil surrounding the root is higher than in the cell. As the water enters the cell, its water potential becomes higher than in the cell next to it, e.g. in the cortex. Therefore, the water moves, by osmosis, into the next cell. The process is repeated until water reaches the xylem. Water also passes from cell to cell along the cell walls.

Mechanism of water movement through a plant

Water vapour evaporating from a leaf creates a kind of suction (**a transpiration pull**) as water molecules are held together by **cohesion**. Therefore, the water forms a **column** and is drawn into the leaf from the xylem. This creates a transpiration stream, pulling water up from the root. Xylem vessels act like tiny tubes, drawing water up the stem by capillary action. Roots also produce a root pressure, forcing water up xylem vessels.
   Refer back to Chapter 2 to remind yourself of the structure of xylem tissue.

Common misconceptions

● Water does not travel through xylem vessels by osmosis. Osmosis involves the movement of water across cell membranes – xylem cells do not have living contents when mature, so there will be no membranes.

44

Wilting

Young plant stems and leaves rely on their cells being turgid to keep them rigid. If the amount of water lost from the leaves of a plant is greater than the amount taken into the roots, the plant will have a water shortage. Cells become flaccid if they lack water, and they will no longer press against each other. Stems and leaves then lose their rigidity and wilt.

Transpiration rate

An increase in temperature increases the kinetic (movement) energy of the water molecules, so they diffuse faster. Transpiration is likely to be faster on a hot day than on a cold day.

An increase in humidity lowers the transpiration rate. This is because it increases the concentration of water molecules outside the leaf, reducing the concentration gradient for diffusion.

Sample question

1 Describe how the structure of xylem tissue is adapted to its functions. [3 marks]

2 Describe the mechanism of water movement through the xylem. [2 marks]

Student's answer

1 The cells join together to make a long ✔ tubular structure. There are no cross-walls ✔ and no living contents, ✔ so the water and mineral salts (✔) can pass through freely.

2 Water moves by the pull from the leaves ✔ caused by transpiration. ✔ Xylem vessels are very thin, so they act like a capillary tube (✔) helping to draw water upwards.

Examiner's comments

Both answers are excellent, gaining the maximum marks available. This candidate has learned the details of water transport in plants really well. The ticks in brackets mean that the statements are correct, but the maximum for the question has already been reached.

Translocation

Translocation is the movement of sugars and amino acids through the phloem sieve tubes of the plant. The **source** is where the materials are produced (usually the leaves) and the **sink** is the region they are transported to. This may be for storage (e.g. in the roots or developing parts of the plants – new leaves, fruits, seeds, etc.), respiration or growth.

During the life of a plant, a region that originally acted as a sink may become a source. For example, sugars stored in the leaves of a bulb in the summer (acting as sink) may be translocated to a growing flower bud or stem the following spring. The bulb is now acting as a source.

Key objectives

The objectives for this chapter are to revise:

- definitions of the key terms
- the circulatory system, including the main blood vessels
- structure of the mammalian heart
- ways of monitoring heart activity
- the effect of physical activity on pulse rate
- coronary heart disease and its possible risk factors
- structures and functions of arteries, veins and capillaries
- components of the blood, red and white blood cells and their functions
- single and double circulatory systems
- functioning of the heart in relation to structure
- the effect of physical activity on heart rate
- prevention and treatment of coronary heart disease
- the process of blood clotting
- transfer of materials between capillaries and tissue fluid

● Transport in animals

The circulatory system is a system of blood vessels with a pump (the heart) and valves to make sure the blood flows one way.

Single circulation

Fish have a heart consisting of one blood-collecting chamber (the atrium) and one blood ejection chamber (the ventricle). It sends blood to the gills, where it is oxygenated. The blood then flows to all parts of the body before returning to the heart. This is known as single circulation because the blood goes through the heart once for each complete circulation of the body.

Double circulation

Blood passes through the heart twice for each complete circulation of the body. The right side of the heart collects deoxygenated blood from the body and pumps it to the lungs. The left side collects oxygenated blood from the lungs and pumps it to the body. The double circulatory system helps to maintain blood pressure, making circulation efficient. Figure 9.1 shows the double circulatory system.

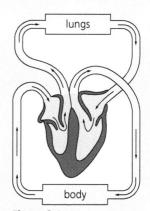

Figure 9.1

● Structure of the heart

The heart is a pump made of muscle that moves blood around the body. The muscle is constantly active, so it needs its own blood supply, through the coronary artery, to provide it with oxygen and glucose.

The heart has two sides. The right side receives deoxygenated blood from the body and then pumps blood to the lungs for oxygenation. The **septum** separates the left side from the right side. The left side receives oxygenated blood from the lungs and pumps it to the body.

Examiner's tips

- To remember which side of the heart contains oxygenated blood, learn this mnemonic: LORD (**L**eft **O**xygenated **R**ight **D**eoxygenated).
- **A**rteries carry blood **A**way from the heart (remember AA).
- On a diagram, the right chambers are on the left of the diagram.

There are four chambers. The right and left atria receive blood from veins and squeeze it into the ventricles. The right and left ventricles receive blood from the atria and squeeze it into arteries. Figure 9.2 shows the main parts of the heart. A surface view of the heart would also show the presence of coronary arteries on the surface of the ventricle muscle walls.

For the Core paper, the valves in the heart do not need to be named. You only need to be able to identify where the valves are and understand that they allow the flow of blood only one way.

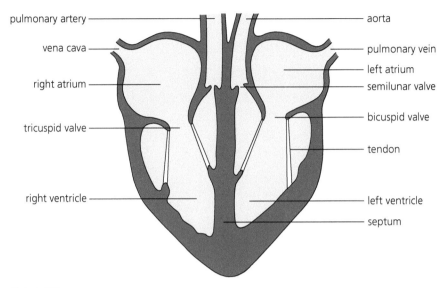

pulmonary artery

vena cava

right atrium

tricuspid valve

right ventricle

aorta

pulmonary vein

left atrium

semilunar valve

bicuspid valve

tendon

left ventricle

septum

Figure 9.2

● Structure of the heart

You need to be able to identify the atrioventricular (bicuspid and tricuspid) and semilunar valves (labelled on Figure 9.2).

The wall of the left ventricle is much thicker than the wall of the right ventricle because it needs to build up enough pressure to move the blood to all of the main organs.

The walls of the atria are much thinner than those of the ventricles. This is because the contraction of the atria needs to be powerful enough only to move blood down into the ventricles, while the ventricles are moving blood around the body and through all of the organs.

The **septum** divides the left side of the heart from the right side. This prevents the mixing of oxygenated and deoxygenated blood.

● Function of the heart

Heart muscles in the atria contract to build up sufficient pressure to move blood through the tricuspid (right side) and bicuspid valves (left side) into the ventricles. These valves then shut to prevent the backflow of blood. As the muscles in the ventricles contract, blood pressure builds up and the blood is forced through the semilunar valves into the pulmonary artery (right side) and aorta (left side). Once the pressure wave has passed, the semilunar valves close to prevent blood from the arteries being sucked back into the ventricles.

> **Now try this**
>
> 1 a On a copy of the diagram of the double circulatory system (Figure 9.1), label:
>   (i)   the four main blood vessels   [4 marks]
>   (ii) the chambers of the heart   [4 marks]
>   (iii) the two valves shown.   [2 marks]
>   b State two differences in composition between blood leaving the right ventricle and
>   blood entering the left atrium.   [2 marks]

Cambridge IGCSE Biology Study and Revision Guide Second Edition © Dave Hayward 2016

Monitoring the activity of the heart

There are a number of ways by which the activity of the heart can be monitored:

- **Pulse rate**: the ripple of pressure that passes down an artery as a result of the heartbeat can be felt as a 'pulse' when the artery is near the surface of the body.
- **Heart sounds** can be heard using a **stethoscope**. This instrument amplifies the sounds of the heart valves opening and closing.
- An **ECG** is an **electrocardiogram**. To obtain an ECG, electrodes, attached to an ECG recording machine, are stuck onto the surface of the skin on the arms, legs and chest. Electrical activity associated with heartbeat is then monitored and viewed on a computer screen or printed out.

> ## Now try this
>
> 2 Trace the diagram of a heart in Figure 9.2 and draw arrows in the four blood vessels and the four chambers to show the direction of blood flow through the heart.
> Shade the right chambers blue to show deoxygenated blood.
> Shade the left chambers red to show oxygenated blood.   [4 marks]

Common misconceptions

- Blood passing through the chambers of the heart does not supply the heart muscle with oxygen or glucose. The heart muscle has its own blood supply – via the coronary arteries – to do this.

Effect of physical activity on pulse rate

A heartbeat is a contraction, each of which squeezes blood to the lungs and body. The heart beats about 70 times a minute, more if you are younger, and the rate becomes lower the fitter you are. This beat can be felt as a pulse in the wrist (radial artery) or neck (carotid artery). During exercise, pulse rate increases from the resting rate and stays high until physical activity slows down or stops. After exercise, the pulse gradually returns to normal.

Effect of physical activity on heart rate

During exercise, heart rate increases to supply the muscles with more oxygen and glucose. These are needed to allow the muscles to respire aerobically, so they have sufficient energy to contract.

Coronary heart disease

Coronary heart disease (a heart attack) is caused by blockage of the coronary arteries. These supply the heart muscle with oxygen and glucose. Without these, the muscle cells stop contracting and die. The possible risk factors are shown in the following table.

Risk factor	Explanation	Preventative measures
Poor diet with too much saturated (animal) fat	Leads to cholesterol building up in arteries, eventually blocking the blood vessel or allowing a blood clot to form	Cholesterol-free diet
Obesity	Being overweight puts extra strain on the heart and makes it more difficult for the person to exercise	Go on a controlled diet and take regular exercise
Smoking	Nicotine damages the heart and blood vessels (see Chapter 15)	Stop smoking
Stress	Tends to increase blood pressure, which can result in fatty materials collecting in the arteries	Find ways of relaxing. Identify the causes of stress and avoid them
Lack of exercise	The heart muscle loses its tone and becomes less efficient in pumping blood	Start taking regular exercise. This increases heart muscle tone, making it more efficient
Genetic predisposition	Heart disease appears to be passed from one generation to the next in some families	Make sure other factors do not increase the risk of heart disease. Monitor health
Age	Risk increases with age	
Sex	Males are more at risk than females	Try to adopt a healthier lifestyle

● Prevention of coronary heart disease

Maintaining a healthy, balanced diet will lower the chance of a person becoming obese. The low intake of saturated fats that is part of a balanced diet reduces the chances of atheroma and thrombus formation. An obese person is less likely to take regular exercise.

Regular, vigorous exercise can also reduce the chances of a heart attack. This may be because it increases muscle tone – not only of skeletal muscle, but also of cardiac muscle. Good heart muscle tone leads to an improved coronary blood flow and the heart requires less effort to keep pumping.

● Treatment of coronary heart disease

The simplest treatment for a patient who suffers from coronary heart disease is to be given a regular dose of aspirin. This prevents the formation of blood clots in the arteries, which can lead to a heart attack.

Methods of removing or treating atheroma and thrombus formations include the use of angioplasty, a stent and, in the most severe cases, bypass surgery.

Angioplasty involves the insertion of a long thin tube called a **catheter** into the blocked or narrowed blood vessel. A wire attached to a deflated balloon is then fed through the catheter to the damaged artery. Once in place, the balloon is inflated to widen the artery wall, effectively freeing the blockage. In some cases, a **stent** is also applied. This is a wire-mesh tube that can be expanded and left in place. It then acts as scaffolding, keeping the blood vessel open and maintaining the free flow of blood. Some stents can give a slow release of chemicals to prevent further blockage of the artery.

In **bypass surgery**, a section of blood vessel from a different part of the body, such as the leg, is removed. This is then attached around the blocked region of artery to bypass it, allowing blood to pass freely. This is a major, invasive operation because it involves open-heart surgery.

● Structures and functions of arteries, veins and capillaries

Figure 9.1 shows all the blood vessels associated with the heart and lungs that you need to learn. Blood vessels associated with the kidneys are shown in Figure 13.1.

Arteries carry blood, at high pressure, away from the heart to organs of the body. **Veins** return blood, at low pressure, from organs towards the heart. **Capillaries** link arteries to veins. They carry blood through organs and tissues, allowing materials to be exchanged.

The following table compares the structures of arteries, veins and capillaries. The explanation of how the structure is related to the function is needed only for the Extended paper.

Blood vessel	Structure	How structure is related to function
Artery	Thick, tough wall with muscles, elastic fibres and fibrous tissue	Carries blood at high pressure – prevents bursting and maintains pressure wave
	Lumen quite narrow, but increases as a pulse of blood passes through	This helps to maintain blood pressure
	Valves absent	The high pressure prevents blood flowing backwards, so valves are not necessary
Vein	Thin wall, mainly fibrous tissue, with little muscle and few elastic fibres	Carries blood at low pressure
	Lumen large	This reduces resistance to blood flow
	Valves present	To prevent backflow of blood
Capillary	Permeable wall, one cell thick, with no muscle or elastic tissue	Allows diffusion of materials between capillary and surrounding tissues
	Lumen approximately one red blood cell wide	White blood cells can squeeze between cells of the wall
	Valves absent	Blood cells pass through slowly to allow diffusion of materials and tissue fluid Blood pressure is lower than in arteries

● Sample question

Figure 9.3 shows a section through the heart.

1 Name the two blood vessels A and B. [2 marks]

2 Which of the blood vessels A, B, C and D carry oxygenated blood? [1 mark]

3 Name valve E and state its function. [3 marks]

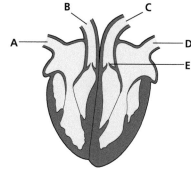

Figure 9.3

Student's answer

1 A, vena cava; ✔ B, pulmonary vein. ✖
2 C ✖
3 Name: semilunar valve; ✔ function: to stop blood going backwards. ✔

Examiner's comments

Blood vessel B is the pulmonary artery. Arteries of the heart always carry blood from a ventricle. Part 2 needs two answers (blood vessels C and D) to gain the mark. Blood vessel D is the pulmonary vein, which carries oxygenated blood to the heart from the lungs. Blood vessel C is the aorta, which carries oxygenated blood from the heart to the body. In part 3, the name of the valve is correct, but there are two marks for its function. This candidate has given only one statement: a second mark was available for stating that the valve prevents blood from going back into the left ventricle.

Cambridge IGCSE Biology Study and Revision Guide Second Edition © Dave Hayward 2016

Functions of arterioles, venules and shunt vessels

The **arterioles**, linking arteries to capillaries, have proportionately more muscle fibres than arteries. When the muscle fibres of the arterioles contract, they make the vessels narrower and restrict the blood flow (a process called **vasoconstriction**; see Chapter 14). In this way, the distribution of blood to different parts of the body can be regulated.

Shunt vessels, linking the arterioles with venules, can dilate to allow the blood to bypass the capillaries in the skin. This helps to reduce heat loss.

Venules return blood to veins.

The lymphatic system

The lymphatic system is a collection of lymph vessels and lymph nodes. It has three main roles:

- The return of tissue fluid to the blood in the form of **lymph fluid**. This prevents fluid build-up in the tissues.
- The production of **lymphocytes**. These are made in lymph glands such as the tonsils, adenoids and spleen. The glands become more active during an infection because they are producing and releasing large numbers of lymphocytes.
- The absorption of **fatty acids** and **glycerol** from the small intestine. Each villus contains a lacteal – a blind-ending lymph vessel.

Blood cells and functions

Blood is made up of a liquid (plasma) containing blood cells. Figure 9.4 gives details of blood cells and their functions. The terms used for naming white blood cells (lymphocytes and phagocytes) are needed only for the Extended paper.

Plasma is a liquid that transports substances to cells and carries wastes away from cells. It acts as a pool for amino acids (these cannot be stored in the body) and contains blood proteins that are important in blood clotting. The following table shows the main substances carried by plasma.

Substance carried in plasma	From	To
Amino acids	Small intestine	Sites of growth and repair
Carbon dioxide	Respiring tissues	Lungs
Glucose	Small intestine	All tissues
Heat	Liver, muscles	All tissues
Hormones, e.g. insulin	Endocrine glands, e.g. pancreas	Target organ, e.g. liver
Urea	Liver	Kidneys

Cambridge IGCSE Biology Study and Revision Guide Second Edition © Dave Hayward 2016

red blood cell

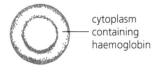

cytoplasm
containing
haemoglobin

biconcave discs with no nucleus –
carry oxygen

lymphocyte

large
nucleus

produce antibodies to fight
bacteria and foreign materials

phagocyte

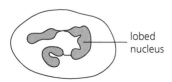

lobed
nucleus

fight disease by surrounding
bacteria and engulfing them

platelets

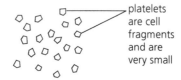

platelets
are cell
fragments
and are
very small

form blood clots, which stop blood
loss at a wound and prevent the
entry of germs into the body

Figure 9.4

● Transport of oxygen

Oxygen is not included in the table above, as it is transported in red blood cells. Oxygen combines with haemoglobin to form oxyhaemoglobin. The oxygen is released from the red blood cells in capillaries where surrounding oxygen levels are low.

● Functions of lymphocytes, phagocytes and platelets

Lymphocytes are involved in the production of antibodies, which are needed to fight disease (see Chapter 10). They can attach themselves to antigens (foreign proteins) and clump them together.

Phagocytes have the ability to change their shape and move to engulf harmful bacteria by a process called **phagocytosis**.

Platelets clump together when tissues are damaged and block the smaller capillaries. The platelets and damaged cells at the wound produce a substance that acts on a soluble plasma protein called **fibrinogen**. As a result, it is changed into insoluble **fibrin**, which forms a network of fibres across the wound. Red blood cells become trapped in this network and so form a blood clot. The clot not only stops further loss of blood, but also prevents the entry of **pathogens** (disease-causing organisms) into the wound.

● Transfer of materials between capillaries and tissue fluid

As blood enters capillaries from arterioles (small arteries), it slows down. This allows substances in the plasma, as well as oxygen from red blood cells, to diffuse through the capillary wall into the surrounding tissues (the capillary wall is thin and permeable). Liquid in the plasma also passes out. This forms tissue fluid, bathing the cells. Waste products from the cells, e.g. carbon dioxide, diffuse back through the capillary walls into the plasma. Some of the tissue fluid also passes back.

Cambridge IGCSE Biology Study and Revision Guide Second Edition   © Dave Hayward 2016

10 Diseases and immunity

Key objectives

The objectives for this chapter are to revise:

- definitions of the key terms
- how the pathogen for a transmissible disease may be transmitted
- the defences of the body and ways of controlling the spread of disease

- how antibodies work and how to explain their specificity
- how active immunity is gained and the process and role of vaccination
- the characteristics and importance of passive immunity
- that some diseases are caused by the immune system targeting and destroying body cells

● Key terms

Pathogen	A disease-causing organism
Transmissible disease	A disease in which the pathogen can be passed from one host to another
Active immunity	The defence against a pathogen by antibody production in the body

● Transmission of pathogens

Pathogens responsible for transmissible diseases can be spread either through direct contact or indirectly.

Direct contact may involve transfer through blood or other body fluids.

- For example, HIV can be passed on by drug addicts who inject the drug into their bloodstream and share needles with other drug users, as the needle will be contaminated.
- Anyone cleaning up dirty needles is at risk of infection if they accidently stab themselves.
- Surgeons carrying out operations have to be especially careful not to be in direct contact with the patient's blood.
- A person with HIV or another sexually transmitted disease (see Chapters 15 and 16) who has unprotected sex can pass on the pathogen to their partner through body fluids.

Indirect contact may involve infection from pathogens on contaminated surfaces, for example during food preparation.

- Raw meat carries bacteria, which are killed if the meat is adequately cooked. However, if the raw meat is prepared on a surface that is then used for other food preparation, e.g. cutting up vegetables that are later eaten raw, the pathogens from the meat can be transferred to the fresh food.
- People handling food are also potential vectors of disease, e.g. if they do not wash their hands after using the toilet.
- Intensive methods of animal rearing may contribute to the spread of infection unless care is taken to reduce the exposure of animals to infected faeces.
- Air-borne infections can be spread by a person with an infection sneezing or coughing. Droplets containing the pathogen float in the air and may be breathed in by other people or fall on to exposed food. Examples of diseases spread in this way include colds, influenza (flu), measles and sore throats.

> **Examiner's tip**
>
> Avoid the term *germ*; always refer to *pathogens* when writing about disease-causing organisms.

● Body defences

The body has three main lines of defence against disease:

- Mechanical barriers
  - The epidermis of the skin is a barrier that prevents bacteria getting into the body.
  - Hairs in the nose help to filter out bacteria that are breathed in.
- Chemical barriers
  - The acid conditions created by hydrochloric acid in the stomach destroy most of the bacteria that may be taken in with food.
  - Mucus, produced by the lining of the trachea and bronchi, is sticky and traps pathogens.
  - Tears contain an enzyme called lysozyme. This dissolves the cell walls of some bacteria, protecting the eyes from infection.
- Cells
  - One type of white blood cell produces antibodies that attach themselves to bacteria, making it easier for other white blood cells to engulf them.
  - Another type of white blood cell engulfs bacteria (a process called phagocytosis) and digests them (see Chapter 9).

● Vaccinations

Vaccination gives a person **immunity** to a specific disease organism, which may otherwise be life threatening if a person is infected by it.

● Common misconceptions

- Pathogens cannot usually be passed on by touching a person with the disease. The pathogen is carried in body fluids such as blood. However, food can be contaminated when a person with pathogens on their skin (e.g. dirty hands) handles it.

● Methods of controlling the spread of disease

- Hygienic food preparation: keeping food-preparation surfaces clean, avoiding the preparation of raw and cooked food on the same surface, cooking food thoroughly to kill any bacteria present.
- Good personal hygiene: washing hands after using the toilet, moving rubbish or handling raw food; avoiding the handling of money when handling unwrapped food.
- Waste disposal: to avoid the development of a breeding ground for pathogens.
- Sewage treatment: to prevent pathogens in faeces from contaminating drinking water and to stop vectors such as flies or rats feeding and transmitting the disease organism.

Cambridge IGCSE Biology Study and Revision Guide Second Edition © Dave Hayward 2016

Action of antibodies

Antibodies are produced by lymphocytes, formed in lymph nodes in response to the presence of pathogens such as bacteria. The pathogens have chemicals called **antigens** on their surface; there is a different antigen with a specific shape for each type of pathogen. So, a different antibody with a matching shape has to be produced for each antigen. The antibodies make bacteria clump together and mark them, in preparation for destruction by phagocytes, or neutralise the toxins produced by the bacteria. Once antibodies have been made, they remain in the blood to provide long-term protection. Some lymphocytes memorise the antigens the body has been exposed to. They can rapidly reproduce and produce antibodies to respond to further infections by the same pathogen.

Now try this

1 Draw a diagram to show how lymphocytes and phagocytes fight pathogens and destroy them. Annotate your diagram. [6 marks]

Active immunity

Some of the lymphocytes that produced the specific antibodies remain in the lymph nodes for some time and divide rapidly and make more antibodies if the same antigen gets into the body again. This creates immunity to the disease caused by the antigen.

Active immunity can also be gained by vaccination.

The process of vaccination

- Vaccination involves a harmless form of the pathogen that has antigens being introduced into the body by injection or swallowing.
- The presence of the antigens triggers lymphocytes to make specific antibodies to combat possible infection.
- Some of these cells remain in the lymph nodes as **memory cells**.
- These can reproduce fast and produce antibodies in response to any subsequent invasion of the body by the same pathogen, providing long-term **immunity**.
- **Mass vaccination** can control the spread of diseases. There needs to be a significant proportion of a population immunised to prevent an epidemic of a disease, ideally over 90%. If mass vaccination fails, the population is at risk of infection with the potential for epidemics.
- **Passive immunity** is a short-term defence against a pathogen.
- It is achieved by injecting the patient with serum taken from a person who has recovered from the disease. Serum is plasma with the fibrinogen removed and contains antibodies against the disease, e.g. tetanus, chickenpox, rabies.
- It is called passive immunity because the antibodies have not been produced by the patient. It is only temporary because it does not result in the formation of memory cells.
- When a mother breastfeeds her baby, the milk contains some of the mother's lymphocytes, which produce antibodies.
- These antibodies provide the baby with protection against infection at a time when the baby's immune responses are not yet fully developed. However, this is another case of passive immunity, as it is only short-term protection: memory cells are not produced.

● Sample question

Distinguish between the terms *active immunity* and *passive immunity*. [4 marks]

Student's answer

Active immunity is a defence against pathogens ✓ by producing antibodies to fight them. ✓ Passive immunity is the same, ✗ but is only short term. ✗

Examiner's comments

The definition of active immunity is correct. However, passive immunity is not the same as active immunity; in passive immunity, the antibodies are acquired from another individual. The statement about passive immunity being short term is correct, but is given as part of a biologically incorrect statement, so it does not gain a mark.

● Autoimmune diseases

Type 1 diabetes mainly affects young people. It is due to the inability of islet cells in the pancreas to produce sufficient insulin. It may be inherited, but can be triggered by an event such as a virus infection. This causes the body's immune system to attack the cells in the pancreas that produce insulin. It is therefore classed as an **autoimmune** disease. As a result, the patient's blood is deficient in insulin and regular injections of the hormone are needed to control blood sugar levels and allow the patient to lead a normal life (see Chapter 14).

Cambridge IGCSE Biology Study and Revision Guide Second Edition © Dave Hayward 2016

Gas exchange in humans

The objectives for this chapter are to revise:

- definitions of the key terms
- the features of gaseous exchange surfaces in humans
- structures associated with the breathing system
- differences in composition between inspired and expired air and the test for carbon dioxide
- the effects of physical activity on rate and depth of breathing
- the functions of cartilage in the trachea
- the role of the ribs, intercostal muscles and diaphragm in ventilation of the lungs
- how to explain differences in composition between inspired and expired air
- how to explain the role of goblet cells, mucus and ciliated cells in protecting the gas exchange system

● Gaseous exchange

This process involves the passage of gases such as oxygen into and carbon dioxide out of cells or a transport system.

First, air needs to be in contact with the gaseous exchange surface. This is achieved by breathing. Figure 11.1 shows the breathing system of a human.

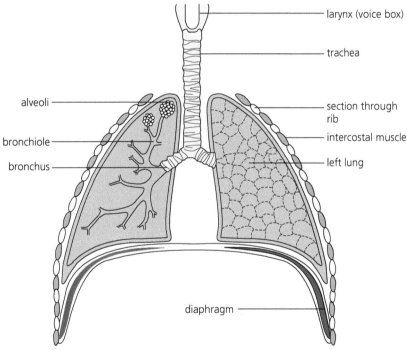

Figure 11.1

Gaseous exchange relies on diffusion. To be efficient, the gaseous exchange surface must:

- be thin – a short distance for gases to diffuse;
- have a large surface area – for gases to diffuse over;
- have good ventilation with air – this creates and maintains a concentration gradient;
- have a good blood supply – to transport oxygen to respiring tissues and bring carbon dioxide from those tissues.

The gaseous exchange surfaces in mammals are the alveoli in the lungs. Figure 11.2 shows the features for gaseous exchange in an alveolus. Figure 11.3 shows the blood supply of the alveoli.

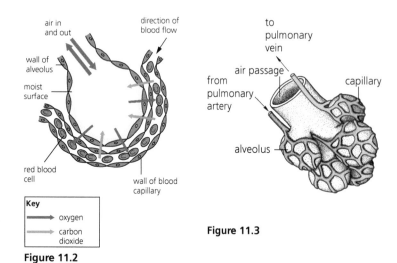

Key
→ oxygen
→ carbon dioxide

Figure 11.2

Figure 11.3

Now try this

1 State how each feature labelled on the diagram of an alveolus (Figure 11.2) makes the process of gaseous exchange efficient. [5 marks]

The composition of inspired and expired air

You need to be able to state the percentages shown in the following table. However, the explanations are needed only for the Extended paper.

Gas	Inspired air/%	Expired air/%	Explanation
Nitrogen	79	79	Not used or produced by body processes
Oxygen	21	16	Used up in the process of respiration, but the system is not very efficient, so only a small proportion of the oxygen available is absorbed from the air
Carbon dioxide	0.04	4	Produced in the process of respiration
Water vapour	Variable	Saturated	Produced in the process of respiration; moisture evaporates from the surface of the alveoli

Testing for carbon dioxide

Limewater can be used to test for carbon dioxide – it changes colour from colourless to milky when the gas is bubbled through.

Expired air makes limewater change colour more quickly than inspired air because there is more carbon dioxide present in expired air.

Effects of physical activity on breathing

Both breathing rate and depth increase during exercise. The volume of air breathed in and out during normal, relaxed breathing is about 0.5 litres. This is the **tidal volume**. Breathing rate at rest is about 12 breaths per minute.

During exercise, the volume inhaled (depth) increases to about 5 litres (depending on the age, sex, size and fitness of the person). The maximum amount of air breathed in or out in one breath is the **vital capacity**. Breathing rate can increase to over 20 breaths per minute.

Now try this

2 a The composition of the air inside the lungs changes during breathing.
   (i) State three differences between inspired and expired air. [3 marks]
   (ii) Gaseous exchange in the alveoli causes some of the changes to the inspired air. Describe three features of the alveoli that assist gaseous exchange. [3 marks]
  b (i) State what is meant by anaerobic respiration. [2 marks]
   (ii) Where does anaerobic respiration occur in humans? [1 mark]

Cambridge IGCSE Biology Study and Revision Guide Second Edition © Dave Hayward 2016

● Sample question

A new, alternative treatment for diabetes is being developed that involves inhaling insulin into the lungs as a spray.

Suggest the path the spray would take from the mouth to enter the alveoli. [3 marks]

Student's answer

The spray would pass through the trachea, then through the bronchioles, then the bronchi to the alveoli. ✓✓

Examiner's comments

The first answer is correct and the candidate has named all the parts involved in the process. However, bronchiole and bronchus are the wrong way round, so only two of the three marks are awarded.

● The role of the ribs, intercostal muscles and diaphragm in ventilation of the lungs

Figure 11.4 shows the relationship between the intercostal muscles, diaphragm and ribcage to achieve ventilation of the lungs.

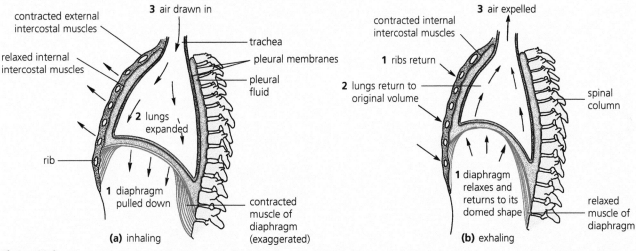

Figure 11.4

Breathing in (inhaling)

The movement of the ribcage is brought about by the contraction of two sets of intercostal muscles that are attached to the ribs. The external intercostal muscles are attached to the external surface of the ribs; the internal intercostal muscles are attached to the internal surface.

When the external intercostal muscles contract, they move the ribcage upwards and outwards, increasing the volume of the thorax.

The diaphragm is a tough, fibrous sheet at the base of the thorax, with muscle around its edge. When the diaphragm muscle contracts, the diaphragm moves down, again increasing the volume of the thorax. This increase in volume reduces the air pressure in the thoracic cavity. As the air pressure outside the body is higher, air rushes into the lungs through the mouth or nose, so ventilation is achieved.

Cambridge IGCSE Biology Study and Revision Guide Second Edition © Dave Hayward 2016

Breathing out (exhaling)

The opposite happens when breathing out; during forced exhalation, the internal intercostal muscles contract and the diaphragm muscles relax. Thoracic volume decreases, so air pressure becomes greater than outside the body. Air rushes out of the lungs to equalise the pressure. Again, ventilation is achieved.

● Common misconceptions

● When you breathe in your ribcage moves up and out (not down and in). Test it out: spread your hands over your chest and take a deep breath in. You can feel your ribcage moving.

> **Now try this**
>
> 3 Write out the series of events involved in breathing in and breathing out as a set of bullet points or as a flowchart linked with arrows. [6 marks]

● Physical activity and the rate and depth of breathing

It has already been stated that the rate and depth of breathing increase during exercise. For limbs to move faster, aerobic respiration in the skeletal muscles increases. Carbon dioxide is a waste product of aerobic respiration (see Chapter 12). As a result, carbon dioxide builds up in the muscle cells and diffuses into the plasma in the bloodstream more rapidly. The brain detects increases in carbon dioxide concentration in the blood and stimulates the breathing mechanism to speed up, increasing the rate of expiration of the gas.

● Protection of the gas exchange system from pathogens and particles

Pathogens, such as bacteria, and dust particles are present in the air we breathe in and are potentially dangerous if not actively removed. Two types of cells (Figure 11.5) provide mechanisms to help achieve this.

Goblet cells are found in the epithelial lining of the trachea, bronchi and some bronchioles of the respiratory tract. Their role is to secrete **mucus**. The mucus forms a thin film over the internal lining. This sticky liquid traps pathogens and small particles, preventing them from entering the alveoli where they could cause infection or physical damage.

Ciliated cells are also present in the epithelial lining of the respiratory tract (see 'Levels of organisation' in Chapter 2). They are in a continually flicking motion to move the mucus, secreted by the goblet cells, upwards and away from the lungs. When the mucus reaches the top of the trachea, it passes down the gullet during normal swallowing.

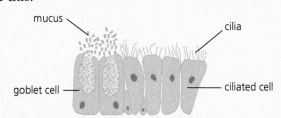

Figure 11.5

12 Respiration

Key objectives

The objectives for this chapter are to revise:

- definitions of the key terms
- the uses of energy in the human body
- that respiration involves the action of enzymes in cells
- the word equations for aerobic and anaerobic respiration in muscles
- the balanced chemical equation for aerobic respiration and for anaerobic respiration in yeast
- the way in which an oxygen debt builds up and how it is removed during recovery

● Key terms

Aerobic respiration	The chemical reactions in cells that use oxygen to break down nutrient molecules to release energy
Anaerobic respiration	The chemical reactions in cells that break down nutrient molecules to release energy without using oxygen

● Respiration

Most of the processes taking place in cells in the body need energy to make them happen. Examples of energy-consuming processes in living organisms are:

- the contraction of muscle cells, e.g. to create movement of the organism;
- synthesis (building up) of proteins from amino acids;
- the process of cell division (Chapter 17) to create more cells, to replace damaged or worn-out cells or to make reproductive cells;
- the process of active transport (Chapter 3), involving the movement of molecules across a cell membrane against a concentration gradient;
- growth of an organism through the formation of new cells or a permanent increase in cell size;
- the conduction of electrical impulses by nerve cells (Chapter 14);
- maintaining a constant body temperature in warm-blooded animals (Chapter 14) to make sure that vital chemical reactions continue at a steady rate even when the surrounding temperature varies.

This energy comes from the food that cells take in. The food mainly used for energy in cells is glucose. Respiration, which produces the energy, is a chemical process that takes place in cells and involves the action of enzymes.

● Common misconceptions

- Respiration is not the same as breathing. Respiration is a process occurring in cells, while breathing is the physical process of ventilating the lungs to obtain oxygen and remove carbon dioxide.

● Aerobic respiration

In humans, energy is usually released by aerobic respiration. However, the cells must receive plenty of oxygen to maintain this process.

The word equation for aerobic respiration is:

glucose + oxygen ⟶ water + carbon dioxide + energy

The breakdown of one glucose molecule releases 2830 kJ of energy.

It is possible to carry out experiments using invertebrates and germinating seeds and measure the oxygen uptake of the organisms: the faster the uptake, the faster the rate of aerobic respiration. Germinating seeds do not need energy for movement, so their respiration rate tends to be lower than that of animals.

Examiner's tips

For the Core curriculum, you need to be able to write the word equations for respiration. The symbol equations are part of the Extended curriculum.

● Anaerobic respiration

This form of respiration does not require oxygen. When tissues are respiring very fast, the oxygen supply is not fast enough to cope, so tissues such as muscles start to respire anaerobically instead. However, this is a less efficient process than aerobic respiration, so much less energy is produced.

The breakdown of one glucose molecule by yeast releases only 118 kJ of energy.

The word equation for anaerobic respiration in yeast is:

glucose → ethanol + carbon dioxide + energy

The word equation for anaerobic respiration in muscles is:

glucose → lactic acid + energy

Examiner's tips

- If you write a symbol equation, you must make sure that the formulae are correct and that the equation is balanced.
- Anaerobic respiration in muscles does not produce carbon dioxide or water.
- Anaerobic respiration in yeast does not produce water.

Now try this

1 Make a table to compare aerobic and anaerobic respiration in humans, with the headings shown below. [6 marks]

Type of respiration	Requirement(s)	Products(s)	Amount of energy released

The balanced chemical equation for aerobic respiration is:

$$C_6H_{12}O_6 + 6O_2 \rightarrow 6H_2O + 6CO_2 + energy$$

It is possible to carry out experiments using germinating seeds to investigate the effect of temperature on respiration rate. Again, the rate of oxygen uptake of the organisms is used to indicate respiration rate. As temperature increases, so does the rate of oxygen uptake and, therefore, the respiration rate. This is because respiration is controlled by enzymes. An increase in temperature increases kinetic energy of the molecules, so reaction rate increases.

The balanced chemical equation for anaerobic respiration in yeast is:

$$C_6H_{12}O_6 \rightarrow 2C_2H_5OH + 2CO_2 + energy$$

Oxygen debt

Muscles respire anaerobically when exercising vigorously because the blood cannot supply enough oxygen to maintain aerobic respiration. However, the formation and build-up of lactic acid in muscles causes cramp (muscle fatigue). An oxygen debt is created because oxygen is needed for aerobic respiration to convert lactic acid back to a harmless chemical (pyruvic acid). This happens in the liver. At the end of a race, a sprinter has to pant to get sufficient oxygen to the muscles to repay the oxygen debt. Breathing remains deep to supply enough oxygen and the heart rate remains fast to transport lactic acid in the blood from the muscles to the liver.

A long-distance runner judges his or her pace, not running too fast, to prevent the muscles respiring anaerobically. Muscle cramp would stop the athlete running.

Sample question

Explain why breathing pattern changes after a period of vigorous exercise. [3 marks]

Student's answer

The breathing rate increases ✓ because muscles build up an oxygen debt ✓ when they respire anaerobically. ✓ Oxygen is needed to break down the lactic acid produced to prevent muscle fatigue.

Examiner's comments

This was an excellent answer, gaining the maximum mark. The last sentence also contained credit-worthy statements.

Cambridge IGCSE Biology Study and Revision Guide Second Edition © Dave Hayward 2016

13 Excretion in humans

Key objectives

The objectives for this chapter are to revise:

- definitions of the key terms
- that urea is formed in the liver from excess amino acids
- the organs responsible for excreting carbon dioxide, urea, water and salts
- that the volume and concentration of urine produced are affected by water uptake, temperature and exercise
- how to identify the ureters, bladder and urethra

- the need for excretion
- the structure of a kidney and the structure and function of a kidney tubule
- dialysis and its application in kidney machines
- the advantages and disadvantages of kidney transplants, compared with dialysis

Key terms

Excretion	The removal from organisms of toxic materials and substances in excess of requirements
Deamination	The removal of the nitrogen-containing part of amino acids to form urea

Examiner's tip

For students following the Extended curriculum, you need to learn a slightly extended definition of excretion, which refers to the removal of waste products of metabolism. You should refer to this in your answer.

Excretion

A number of organs in the body are involved in excretion. These include:

- the liver, which forms urea through the breakdown of excess amino acids (see Chapter 7);
- the lungs, which remove carbon dioxide from the blood (see Chapter 11);
- the kidneys, which filter blood, removing urea, excess water and salts.

The kidneys

Figure 13.1 shows the relative positions of the kidneys, ureters, bladder and urethra in the body. Filtered blood returns to the vena cava (main vein) via a renal vein. The urine formed in the kidney passes down a ureter into the bladder, where it is stored. A sphincter muscle controls the release of urine through the urethra.

The liquid produced by the filtration of blood in the kidneys is called urine – a solution of urea and mineral salts in water. The relative amount of water reabsorbed depends on the state of hydration of the body (how much water is in the blood).

The volume and concentration of urine produced are affected by:

- Temperature: on a hot day, when we sweat more to cool down, the body needs to conserve water, so a small amount of concentrated urine is produced. On a cold day, little sweat is produced, so we tend to produce a larger volume of dilute urine.

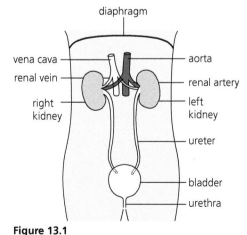

Figure 13.1

Cambridge IGCSE Biology Study and Revision Guide Second Edition © Dave Hayward 2016

- Exercise: more water is lost from the body because of increased sweating, so less is excreted via the kidneys. The release of more sweat helps to maintain the normal body temperature.
- Water intake: the more water taken in, the more urine will be produced and it will become more dilute. If you drink less and other conditions do not change (exercise or external temperature), urine production will decrease and its concentration will increase.

Figure 13.2 shows the water balance of the body. The term *metabolism* refers to chemical processes in cells.

Examiner's tip
Make sure you can label the diagram (Figure 13.1) showing the relative positions of the kidneys, ureters, bladder and urethra. The spellings of ureter and urethra are really important. Check that you get these spellings right, and that the structures are labelled in the correct positions.

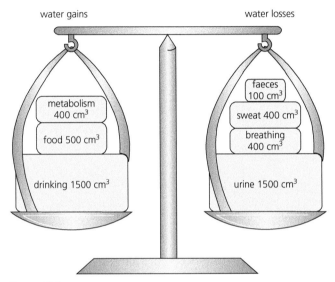

water gains   water losses

metabolism 400 cm³

food 500 cm³

drinking 1500 cm³

faeces 100 cm³

sweat 400 cm³

breathing 400 cm³

urine 1500 cm³

Figure 13.2

Common misconceptions

- Faeces are not an example of excretion – faeces are mainly undigested material that has passed through the gut, but which has not been made in the body. The only excretory materials in it are bile pigments.

Now try this

1 Figure 13.3 shows the human urinary system.

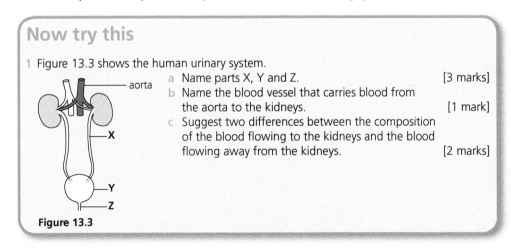

— aorta

— X

— Y

— Z

Figure 13.3

a Name parts X, Y and Z. [3 marks]
b Name the blood vessel that carries blood from the aorta to the kidneys. [1 mark]
c Suggest two differences between the composition of the blood flowing to the kidneys and the blood flowing away from the kidneys. [2 marks]

The role of the liver

Surplus amino acids in the bloodstream cannot be stored. They are removed by the liver and some are assimilated by converting them into proteins. Examples of these include plasma proteins (e.g. fibrinogen in the blood; see Chapter 9). The surplus amino acids are broken down into urea (which is the nitrogen-containing part of the amino acid) and a sugar residue (which can be respired to release energy). The breakdown of amino acids is called **deamination**. Urea is returned to the bloodstream (into the hepatic vein) and filtered out when it reaches the kidneys.

The need for excretion

Some waste products can be toxic. Examples of these include urea and carbon dioxide.

Urea is not normally harmful unless its concentration in the blood gets too high (e.g. if the kidneys fail to excrete it), in which case it starts to be converted into ammonia – a strong, toxic alkali.

Most carbon dioxide is converted to hydrogencarbonate ions (HCO_3^-) and hydrogen ions (H^+) in the blood plasma. A build-up of H^+ will make the blood more acidic. This does not normally happen because of the excretion of carbon dioxide by the lungs.

Structure and function of the kidney

Figure 13.4 shows the structure of a kidney, with its related vessels. Figure 13.5 shows the structure of a kidney tubule.

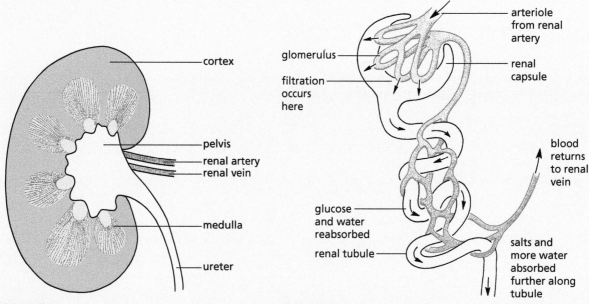

Figure 13.5

Figure 13.4

The kidneys receive blood from the aorta (the main artery) via renal arteries. In the cortex, the renal artery splits into millions of capillaries. Each capillary forms a knotted glomerulus, from which the blood is filtered under pressure. This forces all the small molecules and ions (such as glucose, urea, water and mineral salts) out of the capillary into a tubule. As the filtrate passes through the tubule, reabsorption takes place. Water is reabsorbed by osmosis, while glucose and mineral salts pass back into the blood by diffusion and active uptake (see Chapter 3). The reduction in water content leads to an increase in the concentration of urea. The remaining solution, called urine, passes down the collecting duct of the tubule, through the pelvis of the kidney into the ureter, then down into the bladder for removal through the urethra.

● Dialysis and its application in kidney machines

Dialysis is a method of removing one or more components from a solution using the process of diffusion. The solution is separated from a bathing liquid by a partially permeable membrane (made of cellulose). The bathing liquid contains none of the components that need to be removed from the solution, so these pass from the solution, through the membrane, into the bathing solution by diffusion. The bathing solution needs to be changed regularly to maintain a concentration gradient.

The principle of dialysis is used in a kidney machine, as shown in Figure 13.6.

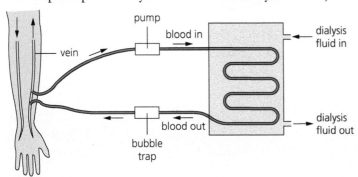

Figure 13.6

A patient with kidney failure needs to have toxic chemicals removed from the blood to stay alive. Blood is removed from a vein in the arm, and is kept moving through dialysis tubing in the dialysis machine using a pump. The tubing is very long to provide a large surface area. The dialysis fluid has a composition similar to blood plasma, but with no urea or uric acid. Urea, uric acid and excess mineral salts are removed from the blood, by diffusion, into the dialysis fluid. The cleaned blood is then passed through a bubble trap to remove any air bubbles, before being returned to the patient's vein.

In summary, the process of dialysis corrects the salt balance of the blood, maintains the glucose concentration and removes urea.

● Advantages and disadvantages of kidney transplants compared with dialysis

Advantages of kidney transplants:

- The patient can return to a normal lifestyle – dialysis may require a lengthy session in hospital, three times a week, leaving the patient very tired after each session.
- A dialysis machine will be available for other patients to use.
- Dialysis machines are expensive to buy and maintain.

Cambridge IGCSE Biology Study and Revision Guide Second Edition © Dave Hayward 2016

Disadvantages of kidney transplants:

- Transplants require a suitable donor with a good tissue match. The donor may be a dead person or a close living relative who is prepared to donate a healthy kidney (humans can survive with one kidney).
- The operation is very expensive.
- There is a risk of rejection of the donated kidney – immunosuppressive drugs have to be used.
- Transplantation is not accepted by some religions.

● Sample question

Outline the function of a renal capsule and renal tubule in the kidney.

[4 marks]

Student's answer

After the blood has been filtered, the capsule collects the filtrate. ✔ This contains water, dissolved salts, glucose and urea. ✔ As the liquid passes along the tubule, all the glucose is reabsorbed, ✔ along with some of the water and salts, ✔ back into the blood capillaries. The remaining filtrate passes down to the ureter.

Examiner's comments

An excellent answer. Marks could also have been gained by referring to the processes involved in the reabsorption process (diffusion and active transport for the glucose; osmosis for the water).

Cambridge IGCSE Biology Study and Revision Guide Second Edition © Dave Hayward 2016

Co-ordination and response

Key objectives

The objectives for this chapter are to revise:

- definitions of the key terms
- nerve impulses and the human nervous system
- how to identify sensory, relay and motor neurones from diagrams
- simple reflex arcs and how to describe a reflex action
- the structures and functions of the parts of the eye and the pupil reflex
- endocrine glands, their secretions and the functions of adrenaline, insulin, oestrogen and testosterone
- structures in the skin
- the maintenance of a constant internal body temperature
- gravitropism and phototropism in shoots and roots
- how to distinguish between voluntary and involuntary actions

- the structure of a synapse, how it works and how it is affected by drugs
- how to explain the pupil reflex and accommodation
- the function of rods and cones
- the role of adrenaline
- the differences between nervous and hormonal control systems
- the concepts of homeostasis and negative feedback, including the maintenance of a constant internal body temperature
- the control of glucose content of the blood
- the symptoms and treatment of Type 1 diabetes
- how to explain gravitropism and phototropism in a shoot and the role of auxin
- the use in weedkillers of 2,4-D

● Key terms

Synapse	A junction between two neurones
Sense organ	A group of receptor cells responding to specific stimuli, such as light, sound, touch, temperature or chemicals
Hormone	A chemical substance produced by a gland and carried in the blood, which alters the activity of one or more specific target organs
Homeostasis	Maintenance of a constant internal environment
Gravitropism	A response in which parts of a plant grow towards or away from gravity
Phototropism	A response in which parts of a plant grow towards or away from the direction from which light is coming

● Nervous control in humans

The human nervous system is responsible for the co-ordination and regulation of body functions. It is made up of two parts:

- central nervous system – brain and spinal cord, the role of which is co-ordination;
- peripheral nervous system – nerves, which connect all parts of the body to the central nervous system.

Sense organs are linked to the peripheral nervous system. They contain groups of receptor cells. When exposed to a stimulus, they generate an **electrical signal** that passes along peripheral nerves to the central nervous system, triggering a response. Nerves contain nerve cells called **neurones**.
  Peripheral nerves contain sensory and motor neurones:

- Sensory neurones transmit nerve impulses from sense organs to the central nervous system.
- Motor (effector) neurones transmit nerve impulses from the central nervous system to effectors (muscles or glands).

Relay (connector) neurones (also called multi-polar neurones) make connections to other neurones inside the central nervous system.

Cambridge IGCSE Biology Study and Revision Guide Second Edition © Dave Hayward 2016

Figure 14.1 shows the structures of neurones. You need to be able to recognise them from their features. Sensory and motor neurones are covered with a myelin sheath, which insulates the neurone to make transmission of the impulse more efficient.

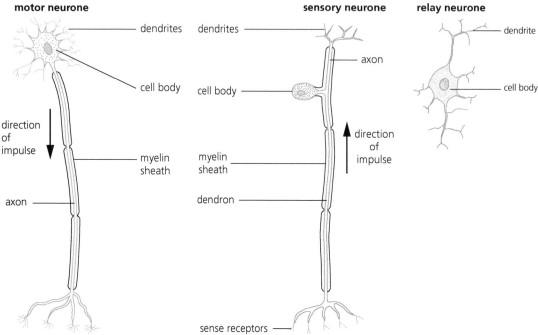

Figure 14.1

The cytoplasm (mainly axon or dendron) is elongated in sensory and motor neurones to transmit the impulse for long distances.

The following table compares the structures of sensory, relay and motor neurones.

Structure	Sensory neurone	Relay neurone	Motor neurone
Cell body	Near the end of the neurone in a ganglion (swelling) just outside the spinal cord	In the centre of the neurone inside the spinal cord	At the start of neurone inside the grey matter of the spinal cord
Dendrites	Present at the end of the neurone	Present at both ends of the neurone	Attached to the cell body
Axon (part of neurone taking impulses away from cell body)	Very short	None	Very long
Dendron	Very long	None	None

● Sample question

Figure 14.2 shows a type of neurone. Name this type of neurone and state a reason for your choice.   [2 marks]

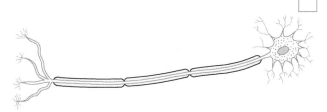

Figure 14.2

Student's answer

Name: motor neurone ✔

Reason: it has a cell body ✘

Examiner's comments

The reason is not enough to distinguish it from other neurones – all neurones have cell bodies. If the answer had been extended to state that the cell body is at the end of the cell, it would have been awarded a mark.

Cambridge IGCSE Biology Study and Revision Guide Second Edition © Dave Hayward 2016

● Simple reflex arcs

A **reflex action** is a means of automatically and rapidly integrating and co-ordinating stimuli with the responses of effectors (muscles or glands). A **reflex arc** describes the pathway of an electrical impulse in response to a stimulus. Figure 14.3 shows a typical reflex arc. The stimulus is a drawing pin sticking in the finger. The response is the withdrawal of the arm caused by contraction of the biceps. Relay neurones are found in the spinal cord, connecting sensory neurones to motor neurones. Neurones do not connect directly with each other; the junction between two neurones is called a **synapse**. The impulse is 'transmitted' across the synapse by means of a chemical.

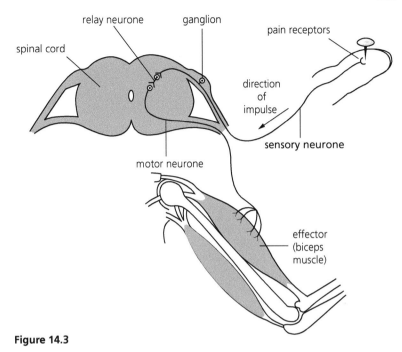

Figure 14.3

The sequence of events is as follows:

Stimulus (sharp pin in finger)

↓

Receptor (pain receptor in skin)

↓

Co-ordinator (spinal cord)

↓

Effector (biceps muscle)

↓

Response (biceps muscle contracts, hand is withdrawn from pin)

Now try this

1 Figure 14.4 shows a nerve cell.
  a (i) Name the type of nerve cell shown in Figure 14.4.   [1 mark]
      (ii) State two features that distinguish it from other types of nerve cell.   [2 marks]
      (iii) Where in the nervous system is this cell located?   [1 mark]
  b Nerve cells are specialised cells. Suggest how the cytoplasm and myelin sheath of the nerve cell, labelled in the figure, enable the nerve cell to function successfully.   [4 marks]

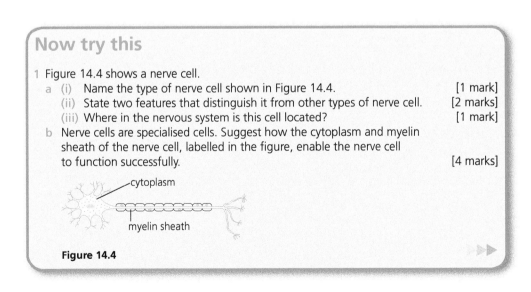

Figure 14.4

Cambridge IGCSE Biology Study and Revision Guide Second Edition © Dave Hayward 2016

c Reflexes involve a response to a stimulus.
  (i) Copy and complete the flowchart below by putting the following terms in the
      boxes to show the correct sequence in a reflex. [2 marks]

co-ordinator   effector   receptor   response   stimulus

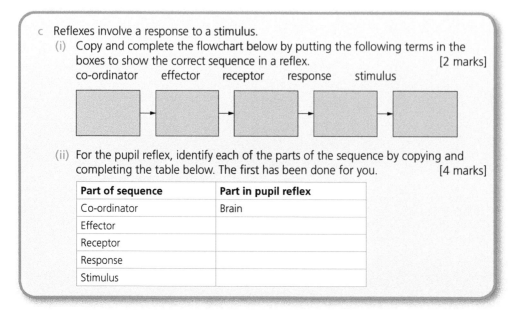

  (ii) For the pupil reflex, identify each of the parts of the sequence by copying and
       completing the table below. The first has been done for you. [4 marks]

Part of sequence	Part in pupil reflex
Co-ordinator	Brain
Effector	
Receptor	
Response	
Stimulus	

Voluntary and involuntary actions

A voluntary action involves the brain in its initiation – it involves conscious thought, as we make a decision about making the action.

Involuntary actions are generally reflexes, which cannot be overridden. They are initiated by sense receptors, which generate electrical impulses. Involuntary actions are automatic, which makes them faster than voluntary actions. Activities inside the body, such as heartbeat and peristalsis, are controlled involuntarily.

The synapse

At a synapse, a branch at the end of one fibre is in close contact with the cell body or dendrite of another neurone (Figure 14.5).

When an impulse arrives at the synapse, **vesicles** in the cytoplasm release a tiny amount of the neurotransmitter substance. It rapidly diffuses across the synaptic gap (also known as the **synaptic cleft**) and binds with **neurotransmitter receptor molecules** in the membrane of the neurone on the other side of the synapse. This then sets off an impulse in the neurone.

Synapses control the direction of impulses because neurotransmitter substances are synthesised on only one side of the synapse, while receptor molecules are present only on the other side. They slow down the speed of nerve impulses slightly because of the time taken for the chemical to diffuse across the **synaptic gap**.

Many drugs produce their effects by interacting with receptor molecules at synapses. **Heroin**, for example, stimulates receptor molecules in synapses in the brain, triggering the release of dopamine (a neurotransmitter), which gives a short-lived 'high'.

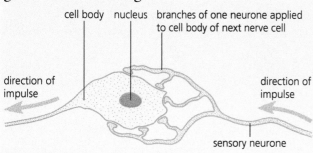

Figure 14.5 Synapses between neurones

Cambridge IGCSE Biology Study and Revision Guide Second Edition © Dave Hayward 2016

● Sense organs

The following table gives examples of sense organs and their stimuli.

Sense organ	Stimulus
Ear	Sound, body movement (balance)
Eye	Light
Nose	Chemicals (smell)
Tongue	Chemicals (taste)
Skin	Temperature, pressure, touch, pain

● Structure and function of parts of the eye

You need to be able to label parts of the eye on diagrams. Figure 14.6 shows the front view of the left eye and Figure 14.7 shows a section through the eye.

The eyebrow stops sweat running down into the eye. Eyelashes help to stop dust blowing on to the eye. Eyelids can close automatically (blinking is a reflex) to prevent dust and other particles getting on to the surface of the cornea. Blinking also helps to keep the surface moist by moving liquid secretions (tears) over the exposed surface. Tears also contain enzymes that have an antibacterial function.

> **Now try this**
>
> 2 Trace or copy both diagrams of the eye (Figures 14.6 and 14.7). Practise adding the labels. [8 marks]

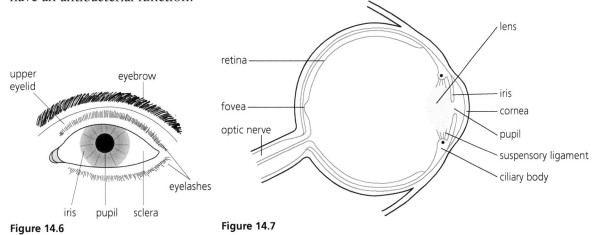

Figure 14.6

Figure 14.7

The following table gives the functions of parts of the eye needed for the Core paper.

Part	Function
Cornea	A transparent layer at the front of the eye that refracts the light entering the eye to help to focus it
Iris	A coloured ring of circular and radial muscle that controls the how much light enters the pupil
Lens	A transparent, convex, flexible, jelly-like structure that focuses light onto the retina
Retina	A light-sensitive layer containing light receptors, some of which are sensitive to the light of different colours
Optic nerve	Transmits electrical impulses from the retina to the brain

● Pupil reflex

This reflex changes the size of the pupil to control the amount of light entering the eye. In bright light, pupil diameter is reduced, as too much light falling on the retina could damage it. In dim light, pupil diameter is increased to allow as much light as possible to enter the eye.

You need to be able to explain the pupil reflex in terms of light intensity and the antagonistic action of the muscles in the iris. **Antagonistic muscles** are those that work in pairs and oppose each other in their actions.

Figure 14.8 shows the effect of light intensity on the iris and pupil.

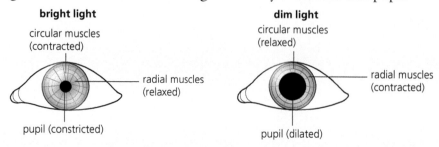

bright light
circular muscles (contracted)
radial muscles (relaxed)
pupil (constricted)

dim light
circular muscles (relaxed)
radial muscles (contracted)
pupil (dilated)

Figure 14.8

The amount of light entering the eye is controlled by altering the size of the pupil.

High light intensity causes a contraction in a ring of **circular muscle** in the iris, while radial muscles relax. This reduces the size of the pupil and reduces the intensity of light entering the eye. High-intensity light can damage the retina, so this reaction has a protective function.

Low light intensity causes circular muscle of the iris to relax and **radial muscle** fibres to contract. This makes the pupil enlarge and allows more light to enter.

The retina detects the brightness of light entering the eye. An impulse passes to the brain along sensory neurones and travels back to the muscles of the iris along motor neurones, triggering a response. The change in the size of the pupil is caused by contraction of the radial or circular muscles.

● Common misconceptions

- Students often confuse circular muscles and ciliary muscles. Circular muscles affect the size of the iris; ciliary muscles affect the shape of the lens.

● Accommodation

The amount of focusing needed by the lens depends on the distance of the object being viewed – light from near objects requires a more convex lens than light from a distant object. The shape of the lens needed to accommodate the image is controlled by the ciliary body – this contains a ring of muscle around the lens.

Distant objects

The ciliary muscles relax, giving them a larger diameter. This pulls on the suspensory ligaments which, in turn, pull on the lens. This makes the lens thinner (less convex). As the ciliary muscles are relaxed, there is no strain on the eye (Figure 14.9, left-hand side).

Near objects

The ciliary muscles contract, giving them a smaller diameter. This removes the tension on the suspensory ligaments which, in turn, stop pulling on the lens. The lens becomes thicker (more convex) (Figure 14.9, right-hand side). As the ciliary muscles are contracted, there is strain on the eye, which can cause a headache if a near object (book, microscope, computer screen, etc.) is viewed for too long.

Cambridge IGCSE Biology Study and Revision Guide Second Edition © Dave Hayward 2016

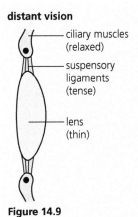

distant vision

ciliary muscles
(relaxed)

suspensory
ligaments
(tense)

lens
(thin)

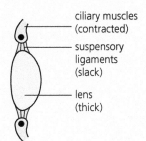

near vision

ciliary muscles
(contracted)

suspensory
ligaments
(slack)

lens
(thick)

Figure 14.9

Now try this

3 Describe and explain how the eye changes its focus from a distant object to a near
object.   [5 marks]

● Rods and cones

Rods and cones are light-sensitive cells in the retina. When stimulated, they
generate electrical impulses, which pass to the brain along the optic nerve.
Cones are most concentrated in the fovea. This is the point on the retina
where the light is usually focused.

The following table shows the main differences between rods and cones.

Cell	Function	Distribution	Comments
Rods	Sensitive to low light intensity. They detect shades of grey	Found throughout the retina, but none in the centre of the fovea or in the blind spot	These cells provide us with night vision, when we can recognise shapes but not colours
Cones	Sensitive only to high light intensity. They detect colour (but do not operate in poor light)	Concentrated in the fovea	There are three types: cells that are sensitive to red light, green light and blue light

● Hormones in humans

Hormones are defined at the start
of this chapter. You need to be able
to identify four endocrine glands
and their secretions. They are
shown in Figure 14.10.

Insulin reduces blood sugar
levels when they get above normal
by instructing the liver to store
them, so removing them from
the blood.

Oestrogen prepares the uterus
for the implantation of the embryo
by making its lining thicker and
increasing its blood supply.

Testosterone plays a part in the
development of male secondary
sexual characteristics.

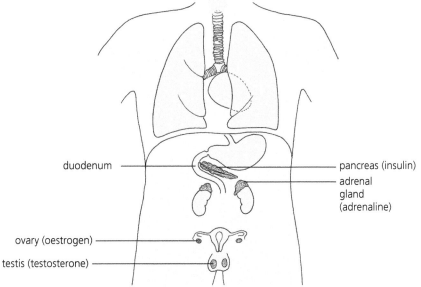

duodenum

pancreas (insulin)

adrenal
gland
(adrenaline)

ovary (oestrogen)

testis (testosterone)

Figure 14.10

Adrenaline

Adrenaline is a hormone secreted into the blood by the adrenal glands, which are found just above each kidney. It causes pulse rate and breathing rate to increase and also widens the pupils. The muscles are prepared for action in a situation such as a fight or struggle, when needing to run away from danger, a sudden shock or a stressful situation, e.g. taking an exam, giving a public performance (hence the expression 'fight or flight').

● Chemical control of metabolic activity by adrenaline

The term **metabolism** describes all the chemical changes that take place in the body.

Adrenaline is a hormone secreted into the blood by the adrenal glands, which are found just above each kidney. It causes pulse rate to increase, so that muscles are supplied with blood containing glucose and oxygen more quickly. This prepares them for action. It also stimulates the liver to convert glycogen to glucose, increasing the blood glucose concentration. Adrenaline also reduces the blood supply to the skin and digestive organs, so blood is diverted to vital organs.

● Comparing nervous and hormonal control systems

The following table shows the main differences between the nervous and hormonal control systems.

Feature	Nervous system	Hormonal (endocrine) system
Form of transmission	Electrical impulses	Chemical (hormones)
Transmission pathway	Nerves	Blood vessels
Speed of transmission	Fast	Slow
Duration of effect	Short term	Long term
Response	Localised	Widespread (although there may be a specific target organ)

Now try this

4 Copy and complete the following table to show the differences between nervous and hormonal control in the human body. [4 marks]

Feature	Nervous control	Hormonal control
Speed	Extremely rapid	
Pathway	Neurones	
Nature of 'impulse'		Chemical
Origin		Endocrine gland

● Homeostasis

Homeostasis is the process of maintaining a constant internal environment, which is vital for an organism to stay healthy. Fluctuations in temperature, water levels and nutrient concentrations, for example, could lead to death.

● Skin structure

You need to be able to name and identify a number of structures in the skin: hairs, erector muscle, sweat gland, receptors (e.g. pressure, temperature, touch, pain), sensory neurone, blood vessels and fatty tissue. These are shown in Figure 14.11.

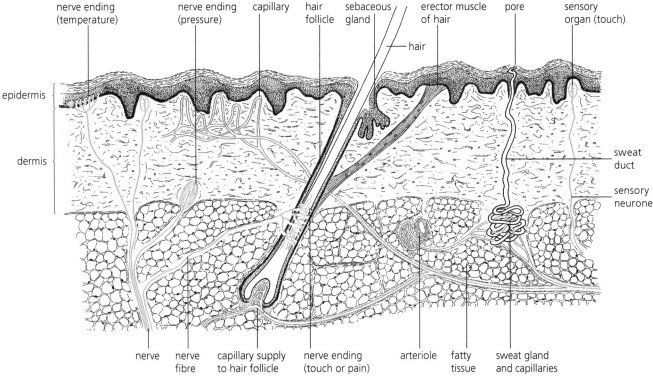

Figure 14.11

● Temperature regulation

Temperature regulation is a homeostatic function. Mammals and birds are warm-blooded – they maintain a constant body temperature despite external environment changes. Humans maintain a body temperature of 37 °C – we have mechanisms to lose heat when we get too hot and ways of retaining heat when we get too cold. Figure 14.12 summarises two ways of regulating body temperature. For the Core exam, you only need to outline the role of sweating, not the role of arterioles.

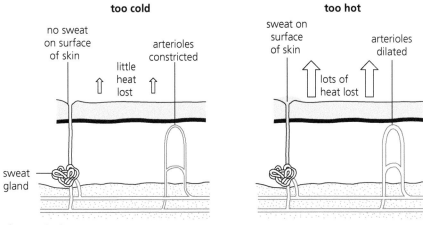

Figure 14.12

Sweating

Sweat is a liquid made up of water, salts and some urea. Sweat glands in the skin secrete sweat through pores on to the skin surface. As the water in the sweat evaporates, it removes heat from the skin, cooling it down. When we are too hot, the volume of sweat produced increases. If we get too cold, the amount of sweat produced is reduced, so less heat is lost through evaporation.

● Common misconceptions

● Sweat does not evaporate from the skin, but water from the sweat evaporates (solutes in the sweat, such as urea and salts, remain on the skin surface).

Insulation

The skin has a layer of fatty tissue that has insulating properties – it reduces heat loss from the skin surface.

Shivering

Uncontrollable bursts of rapid muscular contraction in the limbs release heat as a result of respiration in the muscles. However, its effectiveness in temperature control is questionable.

Role of the brain

The brain plays a direct role in detecting any changes from normal by monitoring the temperature of the blood. A region called the **hypothalamus** contains a thermoregulatory centre in which temperature receptors detect temperature changes in the blood and co-ordinate a response to them. Temperature receptors are also present in the skin. They send information to the brain about temperature changes.

● Role of negative feedback in homeostasis

Homeostasis is the control of internal conditions within set limits. A change from normal, for instance an increase in temperature, triggers a sensor, which stimulates a response in an effector. However, the response – in this case an increase in sweating and vasodilation of arterioles – would eventually result in temperature levels dropping below normal. As temperature levels drop, the sensor detects the drop and instructs an effector (the skin) to reduce sweating and reduce vasodilation of arterioles. This is negative feedback – the change is fed back to the effector.

● Control of glucose content in blood

The liver is a homeostatic organ – it controls the levels of a number of materials in the blood, including glucose. Two hormones – insulin and glucagon – control blood glucose levels. Both hormones are secreted by the pancreas and are transported to the liver in the bloodstream. Excess glucose is stored in the liver and muscles as the polysaccharide glycogen (animal starch). When glucose levels drop below normal, glycogen is broken down to glucose, which is released into the bloodstream.

Cambridge IGCSE Biology Study and Revision Guide Second Edition © Dave Hayward 2016

blood glucose
levels too high

glucose $\xrightarrow{\text{insulin}}$ glycogen

glucagon

blood glucose
levels too low

Examiner's tip

Be accurate with the spellings of **glycogen** and **glucagon** and make sure that you do not get these terms confused.

> **Now try this**
>
> 5 Copy and complete the paragraph using some of the words given below.
>   excretion glucose glycogen insulin liver oestrogen pancreas
>   secretion starch stomach sucrose
>   The bloodstream transports a sugar called _____. The blood sugar level has to be kept constant in the body. If this level falls below normal, a hormone called glucagon is released into the blood by an endocrine organ called the _____. The release of a substance from a gland is called _____. Glucagon promotes the breakdown of _____ to increase the blood sugar level. If the blood sugar level gets too high, the endocrine organ secretes another hormone called _____ into the blood. This hormone promotes the removal of sugar from the blood and its conversion to glycogen in the _____.   [6 marks]

● Type 1 diabetes

There are two types of diabetes. Type 1 is the less common form, the cause of which is outlined in Chapter 10. It results from a failure of cells of the pancreas to produce sufficient insulin. The outcome is that the patient's blood is deficient in insulin. This form of the disease is, therefore, sometimes called 'insulin-dependent' diabetes. The patient is unable to regulate the level of glucose in the blood.

Symptoms include feeling tired, feeling very thirsty, frequent urination and weight loss. Weight loss is experienced because the body starts to break down muscle and fat. Blood glucose may rise to such a high level that it is excreted in the urine, or may fall so low that the brain cells cannot work properly and the person goes into a coma.

Treatment: diabetics need a carefully regulated diet to keep blood sugar within reasonable limits. They also need to take regular exercise and have regular blood tests to monitor their blood sugar levels. The patient needs regular injections of insulin to control blood sugar level and thus lead a normal life.

● Vasodilation and vasoconstriction

Heat is transported around the body in the bloodstream (see Chapter 9). When blood passes through blood vessels near the skin surface, heat is lost by radiation. Arterioles (small arteries) have muscle in their walls.

When we are too hot, these muscles relax, creating a wide lumen through which lots of blood can pass to the skin surface capillaries (the skin of a hot person may look red). This is called **vasodilation**. More heat is radiated, so we cool down.

When we are too cold, the muscles contract, creating a narrow lumen through which little blood can pass (the skin of a cold person may look very pale). This is called **vasoconstriction**. Less heat is radiated to conserve heat.

● Common misconceptions

● The processes of vasodilation and vasoconstriction do not take place in capillaries or veins. They happen only in arterioles. When writing about the processes, make sure you refer to arterioles.

● Tropic responses

Investigations can be carried out to study the effects of gravitropism and phototropism in shoots and roots of plants. Young seedlings and germinating seeds make good subjects for investigations, because they are cheap, easy to prepare and show results quickly. Shoots grow towards light and away from gravity (upwards). Roots grow away from light and towards gravity (downwards).

● Control of plant growth by auxins

Auxins are plant growth substances. They are sometimes referred to as hormones, but this is not very accurate because they are not secreted by glands and are not transported in blood. They are produced by the shoot and root tips of growing plants. An accumulation of auxin in a shoot stimulates cell growth by the absorption of water. However, auxins have the opposite effect in roots – when they build up, they slow down cell growth.

● Phototropism and gravitropism

You need to be able to describe the role of auxins in these processes.

Light

When a shoot is exposed to light from one side, auxins that have been produced by the tip move towards the shaded side of the shoot (or the auxins are destroyed on the light side, causing an unequal distribution). Cells on the shaded side are stimulated to absorb more water than those on the light side, so the unequal growth causes the stem to bend towards the light. Growth of a shoot towards light is called **positive phototropism**.

If a root is exposed to light in the absence of gravity, auxins that have been produced by the tip move towards the shaded side of the root. Cells on the shaded side are stimulated to absorb less water than those on the light side, so the unequal elongation causes the root to bend away from the light. Growth of a root away from light is called **negative phototropism**.

Gravity

Shoots and roots also respond to gravity. If a shoot is placed horizontally in the absence of light, auxins accumulate on the lower side of the shoot, owing to gravity. This makes the cells on the lower side elongate more quickly than those on the upper side, so the shoot bends upwards. This is called **negative gravitropism**.

If a root is placed horizontally in the absence of light, auxins accumulate on the lower side of the root, owing to gravity. However, this makes the cells on the lower side elongate more slowly than those on the upper side, so the root bends downwards. This is called **positive gravitropism**.

Shoots and roots that have their tips removed will not respond to light or gravity because the part that produces auxins has been cut off. Shoots that have their tips covered with opaque material grow straight upwards when exposed to one-sided light because the auxin distribution is not influenced by the light.

Cambridge IGCSE Biology Study and Revision Guide Second Edition © Dave Hayward 2016

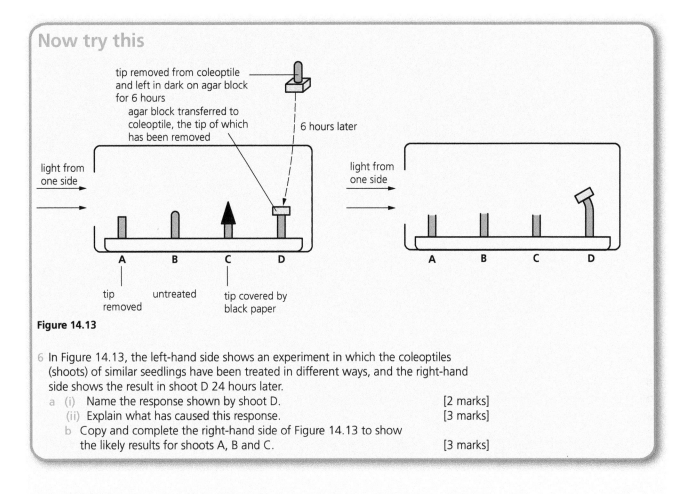

Figure 14.13

6 In Figure 14.13, the left-hand side shows an experiment in which the coleoptiles (shoots) of similar seedlings have been treated in different ways, and the right-hand side shows the result in shoot D 24 hours later.
  a (i) Name the response shown by shoot D.   [2 marks]
      (ii) Explain what has caused this response.   [3 marks]
    b Copy and complete the right-hand side of Figure 14.13 to show the likely results for shoots A, B and C.   [3 marks]

● Effects of synthetic plant hormones used as weedkillers

Synthetic plant hormones are chemicals, similar to auxins, that have been manufactured. If they are sprayed on to plants, they can cause rapid, uncontrolled growth and respiration, resulting in the death of the plant. One effective weedkiller is the synthetic plant hormone 2,4-D. When sprayed on a lawn, it affects the broad-leaved weeds (e.g. daisies and dandelions) but not the grasses. (It is called a 'selective weedkiller'.) Among other effects, it distorts the growth of weeds and speeds up their rate of respiration so much that they exhaust their food reserves and die. Another term for a weedkiller is a **herbicide**.

15 Drugs

Key objectives

The objectives for this chapter are to revise:

- definitions of the key terms
- the use of antibiotics and that some bacteria are resistant to antibiotics
- that antibiotics kill bacteria but do not affect viruses
- the effects of excessive alcohol consumption and the abuse of heroin
- the possible consequences of tobacco smoking, excessive alcohol consumption and injecting heroin

- that the liver is the site of breakdown of alcohol and other toxins
- how the development of resistant bacteria can be minimised
- how heroin affects the nervous system
- the evidence for the link between smoking and lung cancer
- the use of hormones to improve sporting performance

● Key term

Drug   Any substance taken into the body that modifies or affects chemical reactions in the body

● Drugs

The definition of a drug is given above. Drugs may be used to treat disease, reduce the sensation of pain or help calm us down. In addition, they may change our mood by affecting the brain.

● Medicinal drugs

Antibiotics destroy bacteria without harming the tissues of the patient. This makes them ideal for treating bacterial infections. Most of the antibiotics we use come from bacteria or fungi that live in the soil. One of the best-known antibiotics is **penicillin**, which is produced by the mould fungus *Penicillium*.

Antibiotics attack bacteria in a variety of ways. Some of them prevent the bacteria from reproducing or even cause them to burst open; some interfere with protein synthesis to stop bacterial growth.

Not all bacteria are killed by antibiotics. Some bacteria can mutate into forms that are resistant to these drugs, reducing the effectiveness of antibiotics.

It is important to note that antibiotics do not affect viruses.

● Resistant bacteria

It is really important that the development of resistant bacteria is minimised, because otherwise antibiotics will become ineffective in the treatment of bacterial infections. MRSA (methicillin-resistant *Staphylococcus aureus*) is one bacterium that is resistant to antibiotics. It is sometimes called a 'superbug'. To reduce the development of resistant bacteria:

- antibiotics should be used only when essential, otherwise there could be a build-up of a resistant strain of bacteria. The drug resistance can be passed from harmless bacteria to pathogens
- a course of antibiotics should always be completed and not used in a diluted form, or bacteria that have been exposed to the antibiotic but not killed may mutate into resistant forms.

Cambridge IGCSE Biology Study and Revision Guide Second Edition   © Dave Hayward 2016

Antibiotics and viral diseases

Antibiotics are not effective against viral diseases. This is because antibiotics work by disrupting structures in bacteria such as cell walls and membranes, or processes associated with protein synthesis and replication of DNA. Viruses have totally different characteristics to bacteria, so antibiotics do not affect them.

Misused drugs

Alcohol and heroin

Depressants have a relaxing effect because they depress the central nervous system. In high doses, they can cause users to sleep or act as an anaesthetic, causing unconsciousness. You need to be able to describe the effects of alcohol and heroin, the dangers of their misuse and the personal and social problems they can cause.

Drug	Effects	Dangers
Alcohol	Small amounts – alcohol can relax the body and create a sense of wellbeing	Slower reaction time makes driving and handling machinery dangerous. Poor judgement may lead to criminal activity and sexual promiscuity
	Larger amounts slow down the transmission of electrical impulses in the brain, so reactions are depressed, co-ordination is impaired and reasoned judgements become difficult	Long-term excessive drinking can lead to addiction (alcoholism). This can lead to financial difficulties and family problems
		Mood swings involving violence
		As the liver removes alcohol from the blood, heavy drinking can lead to liver damage such as cirrhosis
		Drinking during pregnancy can damage the fetus, increase the risk of miscarriage or premature birth, and reduce average birth weight
Heroin	Heroin is a narcotic, producing a dream-like feeling of relaxation and reducing severe pain. However, it is very addictive, leading to dependency (addiction). Withdrawal symptoms can be very unpleasant, involving cramp, sleeplessness, violent vomiting, sweating and hallucinations	The body develops a tolerance to the drug, so an addict needs to take increasing amounts to achieve the same feeling. This leads to the risk of overdosing on the drug
		When injected using unsterilised and shared needles, there is a risk of infections such as hepatitis and HIV
		Addiction creates financial problems leading to family breakdown, criminal activity and sexual promiscuity

Now try this

1 a Alcohol is described as a depressant and an addictive drug that can damage the body.
   (i) State what is meant by the terms *depressant* and *addictive*. [2 marks]
   (ii) State two long-term effects that alcohol might have on the body. [2 marks]
  b Suggest how alcohol might affect the performance of a car driver. [2 marks]

Examiner's tip

- Before writing an extended answer (usually one that has the instruction 'describe' or 'explain'), look at the value of the question, given in square brackets, e.g. [3 marks]. Plan your answer to include that number of relevant points.
- Avoid writing extra information that you are not sure of because biologically incorrect statements can result in marks being deducted.

● Effects of tobacco smoking

Tobacco smoke contains a large number of toxic chemicals. The main ones are carbon monoxide, nicotine, smoke particles and tar. Note that these chemicals can affect various parts of the body, including the respiratory system and the circulatory system.

Chemical	Effects on respiratory system	Effects on other systems
Carbon monoxide	A poisonous gas. It combines with haemoglobin in red blood cells, preventing them from transporting oxygen	Increases the risk of atherosclerosis and thrombosis, which can lead to coronary heart disease
Nicotine	Addictive, resulting in the continuation of smoking, exposing the lungs to harmful chemicals	• Raises blood pressure and heart rate • Causes thrombosis and can lead to a stroke • Stimulates the brain • Can pass to the blood of a fetus from its mother, resulting in reduced birth weight
Tar	A carcinogen. It increases the risk of lung cancer (cells start to divide out of control). It lines the air passages, increasing mucus production, and paralyses and damages cilia, causing bronchitis	

● Chronic obstructive pulmonary disease (COPD)

This term covers a number of lung diseases, including chronic bronchitis, emphysema and chronic obstructive airways disease. A person suffering from COPD will experience difficulties with breathing, mainly because of narrowing of the airways (bronchi and bronchioles). Symptoms of COPD include breathlessness when active, frequent chest infections and a persistent cough with phlegm (sticky mucus).

● Common misconceptions

• Tar and smoke particles do not enter the blood. Only nicotine and carbon monoxide enter the blood. Tar and smoke particles stay in the lungs.

● Sample question

The following table shows the percentage of haemoglobin that is inactivated by carbon monoxide present in the blood of taxi drivers in a city.

City taxi drivers		Percentage of haemoglobin inactivated by carbon monoxide
Day-time drivers	Non-smokers	2.3
	Smokers	5.8
Night-time drivers	Non-smokers	1.0
	Smokers	4.4

1 Suggest two sources of the carbon monoxide inhaled by these taxi drivers.
[2 marks]

2 Some day-time drivers have 5.8% of their haemoglobin inactivated. Using information from the table, explain which source contributes most to his effect. [2 marks]

3 Suggest a reason for the differences, shown in the table, between day-time and night-time drivers. [1 mark]

Student's answer

1 (1) Cigarette smoke ✓
  (2) Breathing by passengers ✗
2 It must be cigarette smoking ✓ because non-smokers have less of their haemoglobin affected. ✓
3 There could be less car exhaust fumes, containing carbon monoxide, at night. ✓

Examiner's comments

In part 1, the second answer is biologically incorrect (we breathe out carbon dioxide, not carbon monoxide). The other correct answer was car exhaust gases. In part 2, the answer and the explanation were correct. Part 3 was a good answer.

● Role of the liver in the breakdown of drugs

The body treats alcohol as a poison. The liver removes poisons, such as alcohol and drugs, from the blood and breaks them down. Prolonged and excessive use of alcohol damages the liver and may cause it to fail. An overdose of drugs, such as paracetamol, can result in death owing to liver failure, because the liver cannot cope with breaking down such a high concentration of the chemical. The liver also converts hormones (some of which are misused as drugs) into inactive compounds. These are filtered out of the blood by the kidneys.

● How heroin affects the nervous system

Heroin produces its effects by interacting with receptor molecules at **synapses** (see Chapter 14). Synapses are tiny gaps between neurones, across which electrical impulses cannot jump. To maintain the transmission of the impulse, a chemical messenger called a neurotransmitter is released into the gap. Heroin mimics the transmitter substances in synapses in the brain, causing the stimulation of receptor molecules. This causes the release of **dopamine** (a neurotransmitter), which gives a short-lived 'high'.

● Evidence for the link between smoking and lung cancer

The fact that a higher risk of dying from lung cancer is correlated with heavy smoking does not actually prove that smoking is the cause of lung cancer, but the weight of evidence supporting the link is now very great. Many scientific studies show, beyond all reasonable doubt, that the large increase in lung cancer is almost entirely a result of cigarette smoking. There are at least 17 substances in tobacco smoke known to cause cancer in experimental animals, and it is now thought that 90% of lung cancer is caused by smoking.

Cambridge IGCSE Biology Study and Revision Guide Second Edition © Dave Hayward 2016

● Performance-enhancing hormones

These are used illegally by some athletes and sports to boost their performance. Some of these drugs are synthetic forms of hormones.

Testosterone is made in the testes of males and is responsible for promoting male primary and secondary sexual characteristics. Taking testosterone supplements (known as 'doping') leads to increased muscle and bone mass. Therefore, the practice has the potential to enhance a sportsperson's performance.

Anabolic steroids are synthetic derivatives of testosterone. They affect protein metabolism, increasing muscle development and reducing body fat. Athletic performance is enhanced as a result. There are serious long-term effects of taking anabolic steroids, including sterility, masculinisation in women, and liver and kidney malfunction.

Cambridge IGCSE Biology Study and Revision Guide Second Edition © Dave Hayward 2016

Key objectives

The objectives for this chapter are to revise:

- definitions of the key terms
- examples of asexual reproduction
- parts of insect-pollinated and wind-pollinated flowers, their functions and adaptations
- pollination and fertilisation
- environmental conditions that affect seed germination
- parts of the male and female reproductive systems and their functions
- adaptive features of sperm and egg cells
- fertilisation in humans
- early development, growth and development of the fetus
- antenatal care of pregnant women and the processes involved in labour and birth
- roles of oestrogen and testosterone in puberty
- the menstrual cycle

- main methods of birth control
- sexually transmitted infections and control of their spread
- advantages and disadvantages of asexual and sexual reproduction
- the terms *haploid* and *diploid*
- *self-pollination* and *cross-pollination* and their implications to a species
- the process leading up to fertilisation in a flower
- functions of the placenta and umbilical cord
- advantages and disadvantages of breastfeeding compared with bottle–feeding
- the production and roles of hormones in the menstrual cycle and in pregnancy
- use of hormones in contraception and fertility treatments
- artificial insemination and *in vitro* fertilisation
- how HIV affects the immune system

● Key terms

Asexual reproduction	The process resulting in the production of genetically identical offspring from one parent
Sexual reproduction	A process involving the fusion of two gametes (sex cells) to form a zygote and the production of offspring that are genetically different from each other
Fertilisation	The fusion of gamete nuclei
Pollination	The transfer of pollen grains from the anther to the stigma
Sexually transmitted infection	An infection that is transmitted via body fluids through sexual contact
Self-pollination	The transfer of pollen grains from the anther of a flower to the stigma of the same flower or a different flower on the same plant
Cross-pollination	The transfer of pollen grains from the anther of a flower to the stigma of a flower on a different plant of the same species

● Asexual reproduction

Examples of organisms that show this form of reproduction include bacteria, fungi and potatoes. These are described in Figure 16.1.

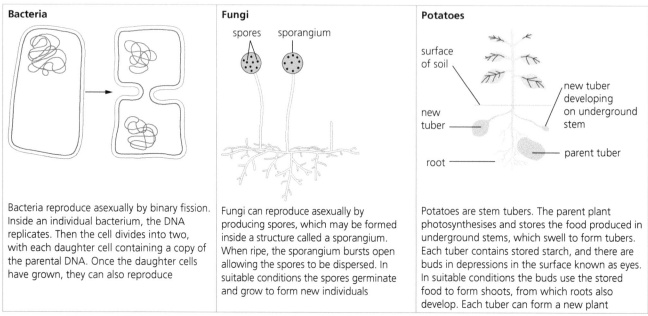

Bacteria

Bacteria reproduce asexually by binary fission. Inside an individual bacterium, the DNA replicates. Then the cell divides into two, with each daughter cell containing a copy of the parental DNA. Once the daughter cells have grown, they can also reproduce

Fungi

spores   sporangium

Fungi can reproduce asexually by producing spores, which may be formed inside a structure called a sporangium. When ripe, the sporangium bursts open allowing the spores to be dispersed. In suitable conditions the spores germinate and grow to form new individuals

Potatoes

surface of soil

new tuber developing on underground stem

new tuber

parent tuber

root

Potatoes are stem tubers. The parent plant photosynthesises and stores the food produced in underground stems, which swell to form tubers. Each tuber contains stored starch, and there are buds in depressions in the surface known as eyes. In suitable conditions the buds use the stored food to form shoots, from which roots also develop. Each tuber can form a new plant

Figure 16.1

Advantages and disadvantages of asexual reproduction

The advantages and disadvantages for a species of asexual reproduction are listed in the table. You need to be able to relate these to a population in the wild and to crop production.

Advantages	Disadvantages
The process is quick	There is little variation created, so adaptation to a changing environment (evolution) is unlikely
Only one parent is needed	If the parent has no resistance to a particular disease, none of the offspring will have resistance
No gametes are needed	Lack of dispersal (e.g. potato tubers) can lead to competition for nutrients, water and light
All the good characteristics of the parent are passed on to the offspring	
Where there is no dispersal (e.g. in potato tubers), offspring will grow in the same favourable environment as the parent	
Plants that reproduce asexually usually store large amounts of food that allow rapid growth when conditions are suitable	

In **agriculture** and **horticulture**, asexual reproduction is exploited to preserve desirable qualities in crops: the good characteristics of the parent are passed on to all the offspring. The bulbs produced can be guaranteed to produce the same shape and colour of flower from one generation to the next. In some cases, such as tissue culture, the young plants grown can be transported much more cheaply than, for example, potato tubers, as the latter are much heavier and more bulky. The growth of new plants by asexual reproduction tends to be a quick process.

In the wild, it might be a disadvantage to have no variation in a species. If the climate or other conditions change and a vegetatively produced plant has no resistance to a particular disease, the whole population could be wiped out.

Cambridge IGCSE Biology Study and Revision Guide Second Edition © Dave Hayward 2016

Sexual reproduction

The definitions of the terms *sexual reproduction* and *fertilisation* are given at the start of this chapter. You need to learn these.

You need to be able to state that the nuclei of gametes (sex cells) are haploid (see Chapter 17) and the nucleus of the zygote (a fertilised egg cell) is diploid.

Advantages and disadvantages of sexual reproduction

Advantages	Disadvantages
There is variation in the offspring, so adaptation to a changing or new environment is likely to occur, enabling survival of the species	Two parents are usually needed (although not always – some plants can self-pollinate)
New varieties can be created, which may have resistance to disease	Growth of a new plant to maturity from a seed is slow
In plants, seeds are produced, which allows dispersal away from the parent plant, reducing competition	

You need to be able to relate these advantages and disadvantages to a population in the wild and to crop production. The table covers some of these points.

Sexual reproduction is exploited in agriculture and horticulture to produce new varieties of animals and plants by cross-breeding. The new varieties can have the combined features of the organisms used to produce them.

Sexual reproduction in plants

You need to be able to describe the structure and functions of parts of an insect-pollinated flower. Figure 16.2 shows the main parts of a lupin flower that has been cut in half. Other flowers have the same features, but the numbers and relative sizes of the parts vary.

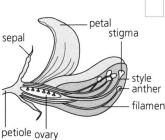

Figure 16.2

Functions of parts of a flower

The following table shows the main functions of the parts of a flower.

Part	Function
Petal	Often large and coloured to attract insects
Sepal	Protects the flower while in bud
Stamen	The male reproductive part of the flower, made up of the anther and filament
Anther	Contains pollen sacs in which pollen grains are formed. Pollen contains male sex cells. Note: you need to be able describe an anther
Filament	Supports the anther
Carpel	The female reproductive part of the flower, made up of the stigma, style and ovary
Stigma	A sticky surface that receives pollen during pollination. Note: you need to be able to describe a stigma
Style	Links the stigma to the ovary through which pollen tubes grow
Ovary	Contains ovules
Ovule	Contains a haploid nucleus, which develops into a seed when fertilised

The definition of *pollination* has been given at the start of this chapter.

Cambridge IGCSE Biology Study and Revision Guide Second Edition © Dave Hayward 2016

Agents of pollination

Flowers are usually pollinated by insects or wind. The structural adaptations of a flower depend on the type of pollination the plant uses.

Insect- and wind-pollinated flowers

Figure 16.3 shows the structure of a grass flower that is wind pollinated. Although you do not need to be able to draw this, you do need to be able to compare insect- and wind-pollinated flowers.

The following table compares the features of wind- and insect-pollinated flowers.

Feature	Insect pollinated	Wind pollinated
Petals	Present – often large, coloured and scented, with guidelines to guide insects into the flower	Absent, or small and inconspicuous
Nectar	Produced by nectaries to attract insects	Absent
Stamen	Present inside the flower	Long filaments, allowing the anthers to hang freely outside the flower so the pollen is exposed to the wind
Stigmas	Small surface area, inside the flower	Large and feathery, hanging outside the flower to catch pollen carried by the wind
Pollen	Smaller numbers of grains – grains are often round and sticky, or covered in spikes to attach to the furry bodies of insects	Larger numbers of smooth and light pollen grains, which are easily carried by the wind
Bracts (modified leaves)	Absent	Sometimes present

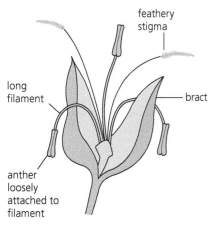

feathery stigma

long filament

bract

anther loosely attached to filament

Figure 16.3

● Common misconceptions

- Pollination and seed dispersal are not the same thing. When animals such as insects carry pollen, they aid pollination. When animals carry seeds, they aid seed dispersal.

Fertilisation

Fertilisation happens when the pollen nucleus fuses with the nucleus of the ovule.

Environmental conditions affecting germination

A seed is a living structure. It contains an embryo that will germinate and develop into an adult plant if provided with suitable conditions. These are listed and explained in the following table.

Cambridge IGCSE Biology Study and Revision Guide Second Edition © Dave Hayward 2016

Environmental condition	Explanation
Water	Water is absorbed through the micropyle until the radicle has forced its way out of the testa. It is needed to activate enzymes that convert insoluble food stores into soluble foods that can be used for growth and energy production
Oxygen	Oxygen is needed for respiration to release energy for growth and the chemical changes needed for mobilisation of food reserves
Suitable temperature	Enzymes work best at an optimum temperature. Generally, the higher the temperature (up to 40 °C), the faster the rate of germination. However, some seeds need a period of chilling before they will germinate. Low temperatures usually maintain dormancy – if the seed germinated in unsuitable conditions, it would be unlikely to survive

Self-pollination and cross-pollination

Self-pollination involves the transfer of pollen from the anther to the stigma of the same flower or to another flower of the same plant. Pollinators are not needed and smaller numbers of pollen grains need to be produced because there is a greater chance of successful pollination. This increases the chance of fertilisation and seed formation, but reduces the variation in the offspring. Self-pollinated plants are less likely to cope with adapting to environmental change.

Cross-pollination involves the transfer of pollen from the anther of a flower to the stigma of a flower on a different plant of the same species. This reduces the chance of fertilisation (wind-pollinated flowers produce large numbers of pollen grains because of the wastage involved), but increases variation and the ability to adapt to environmental change. In addition, pollinators are needed for this process.

Growth of pollen tube and the process of fertilisation

Figure 16.4 shows a section through a single carpel. If pollen grains are of the same species as the flower they land on, they may germinate. Germination is triggered by a sugary solution on the stigma and involves the growth of a pollen tube from the pollen grain. The pollen tube contains the male nucleus, which is needed to fertilise the ovule inside the ovary. The pollen tube grows down the style, through the ovary wall and through the micropyle of the ovule. Fertilisation is the fusion of the male nucleus (in the pollen grain) with the female nucleus (in the ovule). If the ovary contains a lot of ovules, each will need to be fertilised by a different pollen nucleus.

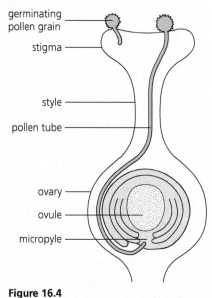

Figure 16.4

● Sexual reproduction in humans

Structure and function of parts of the male reproductive system

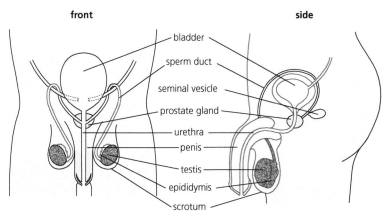

Figure 16.5

Part	Function
Penis	Can become firm so that it can be inserted into the vagina of the female during sexual intercourse to transfer sperm
Prostate gland	Adds fluid and nutrients to sperm to form semen
Scrotum	A sac that holds the testes outside the body, keeping them cooler than body temperature
Sperm duct	Muscular tube that links the testis to the urethra to allow the passage of semen containing sperm
Testis	Male gonads that produce sperm
Urethra	Passes semen containing sperm to the penis; also carries urine from the bladder at different times

Structure and function of parts of the female reproductive system

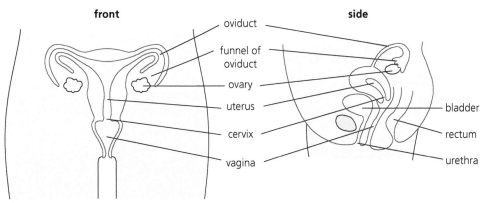

Figure 16.6

Part	Function
Cervix	A ring of muscle that separates the vagina from the uterus
Ovary	Contains follicles in which ova (eggs) are produced
Oviduct	Carries an ovum to the uterus, with propulsion provided by tiny cilia in the wall; also the site of fertilisation
Uterus	Where the fetus develops
Vagina	Receives the male penis during sexual intercourse; sperm are deposited here

● Common misconceptions

- Terms such as **urethra** and **ureter** are often confused or misspelt. Make sure you can spell and label these parts correctly.

Cambridge IGCSE Biology Study and Revision Guide Second Edition © Dave Hayward 2016

Adaptive features of sperm and egg cells

Diagrams of sperm and egg cells can be found in Figure 2.4.

Sperm cells have a flagellum (tail) to swim with and the tip of the sperm (called an acrosome) produces enzymes to digest the cells around an egg and the egg membrane.

Egg cells have a relatively large amount of cytoplasm in which there are energy stores (yolk droplets containing fat). They have a jelly coat that changes after fertilisation.

Comparing sperm and egg cells

Adaptive features of sperm cell	Adaptive features of egg cell
Motile – flagellum so that the cell can swim to the egg from the vagina to the oviduct	Not motile
Size – much smaller than an egg cell	Size – much larger than a sperm cell
Mitochondria present in the mid-piece to provide energy for movement	There are energy stores (yolk droplets with fat) in the cytoplasm
Numbers – millions present in a single ejaculation	Numbers – one egg released each month
Acrosome secretes enzymes to digest the cells around an egg and the egg membrane	Jelly coat acts as a barrier to sperm cells of the wrong species. It binds with a sperm cell during fertilisation and triggers enzymes to be released from the sperm's acrosome; after fertilisation it breaks down

Fertilisation

Fertilisation is the fusion of the nuclei from a male gamete (sperm) and a female gamete (egg cell/ovum). It occurs in the oviduct.

Formation and development of the fetus

The zygote starts to divide by mitosis to form a ball of cells (a blastula). It continues to move down the oviduct until it reaches the uterus.

Implantation occurs when the blastula embeds in the lining of the uterus.

Development of the fetus – the blastula develops into an embryo and some of the cells form a placenta, linking the embryo with the uterus lining. Organs such as the heart develop and, after eight weeks, the embryo is called a fetus. Growth of the fetus requires a good supply of nutrients and oxygen. This is achieved through the link between the placenta and the mother's blood supply in the uterus lining (see Figure 16.7). The **placenta** transfers oxygen to the blood supply of the fetus, which passes through the **umbilical cord** and removes carbon dioxide and other waste products. The fetus is surrounded by an amniotic sac. The **amniotic sac** is a membrane, formed from cells of the embryo, which contains the amniotic fluid. It encloses the developing fetus and prevents the entry of bacteria.

Amniotic fluid supports the fetus, protecting it from physical damage. It absorbs excretory materials (urine) released by the fetus.

As the fetus grows in the early stages it becomes increasingly complex, with systems of the body developing. Towards the end of pregnancy, its size increases substantially.

Cambridge IGCSE Biology Study and Revision Guide Second Edition © Dave Hayward 2016

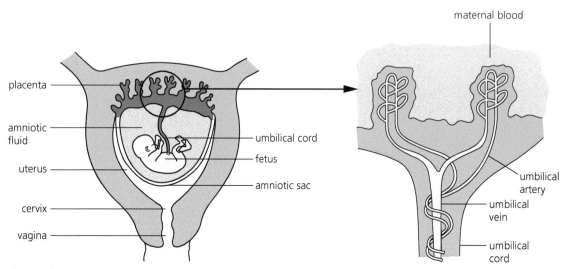

Figure 16.7

Antenatal care

Before the baby is born, it obtains all its dietary requirements from its mother through the placenta. The mother's diet needs to be balanced so that the fetus receives all the materials needed for healthy growth and development. Amino acids are needed to form proteins for growth; calcium is needed for the development of the skeleton; iron is needed for red blood cell formation. If the mother's diet is deficient in any nutrient, the baby may not develop properly. This means that the mother needs more protein, calcium, iron, vitamin C and energy (from carbohydrates or fats) than she usually needs in her diet.

There is a risk to the fetus if the mother smokes or drinks excessively: it may have a lower than average birth weight and may be more prone to illness. Exposure of the fetus to alcohol can also cause a permanent condition called fetal alcohol syndrome.

Birth

- The first stage of the birth process is called labour.
- The muscular walls of the uterus start to contract.
- The pressure breaks the amniotic sac, releasing the amniotic fluid (this is known as the waters breaking).
- Contractions gradually become more frequent, pushing the baby down towards the cervix, which becomes dilated to allow the baby to pass through.
- The vagina stretches to allow the baby to be born.
- The baby is still attached to the placenta by the umbilical cord, so this is cut and tied.
- The placenta breaks away from the wall of the uterus and passes out (this is known as the afterbirth).

> **Now try this**
>
> 1 Describe, in sequence, the main events that occur during birth.   [3 marks]

Function of the placenta and umbilical cord

The placenta brings the blood supply of the fetus close to that of the mother, but prevents mixing. This is really important because the fetus and mother may have different blood groups – any mixing could result in blood clotting, which could be fatal to both mother and fetus. Blood from the fetus passes through the umbilical cord, in the umbilical artery, to the placenta. Here it comes close to the mother's blood. Oxygen, amino acids, glucose and other nutrients diffuse into the blood of the fetus from the mother's blood. Carbon dioxide, urea and other wastes pass into the mother's blood from the blood of the fetus. Blood returns to the fetus through the umbilical vein, also in the umbilical cord. The placenta acts as a barrier to toxins and pathogens.

However, some drugs (such as aspirin and heroin), along with nicotine and carbon monoxide from smoking, alcohol from drinks and viruses such as HIV and rubella (German measles), can all pass across the placenta, risking the health of the developing fetus.

Now try this

2 Figure 16.8 shows a fetus developing in the uterus.
  a Copy the figure and label parts A and B.   [2 marks]
  b Outline three functions of the placenta.   [3 marks]
  c The blood of the fetus and that of the mother flow close to each other in the placenta, but do not mix. State two advantages to the fetus of having a separate blood system from that of the mother.   [2 marks]

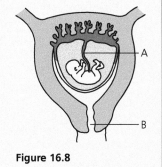

Figure 16.8

● Sample question

Figure 16.9 shows a fetus developing in the uterus. Copy and complete the table below by identifying the parts labelled A, B and C and stating a function of each one.   [6 marks]

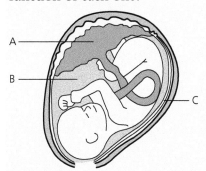

Figure 16.9

Part	Name	Function
A		
B		
C		

Student's answer

Part	Name	Function
A	Placenta ✔	Provides the fetus with blood containing oxygen from the mother ✘
B	Amniotic fluid ✔	Protects the fetus ✘
C	Uterus ✘	Contains the fetus during pregnancy ✘

Examiner's comments

The description of the function of the placenta is very badly worded: the placenta prevents the blood of the mother and fetus from mixing. Answers containing biologically incorrect information are penalised.

Details about the amniotic fluid are too vague to gain the mark for the function. The correct answer was 'to protect the fetus from physical damage'.

Part C is the amniotic sac, which contains the amniotic fluid.

Advantages and disadvantages of breastfeeding over bottle-feeding

Advantages (some are to the baby, while others are to the mother):

- There are antibodies present in breast milk, giving the baby protection against infection.
- Foodstuffs are present in breast milk in the correct proportions, with no additives or preservatives.
- There is no risk of an allergic reaction to breast milk.
- Breast milk is produced at the correct temperature.
- Breastfeeding builds a bond between mother and baby.
- Breast milk does not require sterilisation, as there are no bacteria present that could cause intestinal disease.
- There is no cost involved in using breast milk and it does not need to be prepared.
- Breastfeeding triggers a reduction in the size of the mother's uterus.

Disadvantages (all are to the mother):

- The mother has to do the feeding (although milk could be expressed into a container).
- The process of breastfeeding can be painful.
- It is not possible to measure how much milk the baby has consumed.
- It is more difficult for a breastfeeding mother to go back to work.
- The mother may need to avoid some strongly flavoured foods in her diet.

> **Examiner's tip**
>
> You are unlikely to need to describe all the advantages and disadvantages of breastfeeding. Choose three or four from the list that are easy to remember.

Sex hormones in humans

These are responsible for the development of secondary sexual characteristics at puberty. Testosterone, secreted by the testes, causes the changes in boys; oestrogen, secreted by the ovaries, causes the changes in girls. Puberty is when the sex organs (ovaries in girls; testes in boys) become mature and start to secrete hormones and make gametes (ova and sperm). Puberty happens usually between the ages of 10 and 14 years, but varies from person to person.

The following table shows the secondary sexual characteristics that appear at puberty. A drop in hormone levels can reduce these features, while a high level of hormone can increase them.

Male	Female
Voice becomes much lower (breaks)	Breasts grow, nipples enlarge
Hair starts to grow on chest, face, under arms and in pubic area	Hair develops under arms and in pubic area
Body becomes more muscular	Hips become wider
Penis becomes larger	Uterus and vagina become larger
Testes start to produce sperm	Ovaries start to release eggs and periods begin (menstruation)

The menstrual cycle

This is a cycle involving changes in the uterus and ovaries, controlled by a number of hormones. Each cycle takes about 28 days:

- At the start of each cycle, menstruation occurs – the lining of the uterus breaks down, and the cells and blood in the lining are shed via the vagina. This is **menstruation**.
- The uterus lining then starts to build up again, developing a mass of blood vessels so that it is ready to receive a fertilised ovum.
- A follicle in one of the ovaries matures into an ovum.
- About half-way through the cycle, the wall of the ovary ruptures and an ovum is released.
- Towards the end of the cycle, the lining of the uterus breaks down again.

Now try this

3 Figure 16.10 represents part of the male reproductive system, together with parts of the urinary system.
   a Copy or trace the figure and label:
      (i) the sperm duct (vas deferens)   [1 mark]
      (ii) the urethra.   [1 mark]
   b What is the difference in function of the urethra between males and females?   [2 marks]
   c (i) The hormone testosterone controls the development of secondary sexual characteristics in males. State two of these characteristics that develop at puberty.   [2 marks]
      (ii) On your drawing, label clearly where this hormone is produced.   [1 mark]
      (iii) Some international athletes, female as well as male, have taken testosterone, illegally, as a drug. Suggest why these athletes might have done this.   [2 marks]

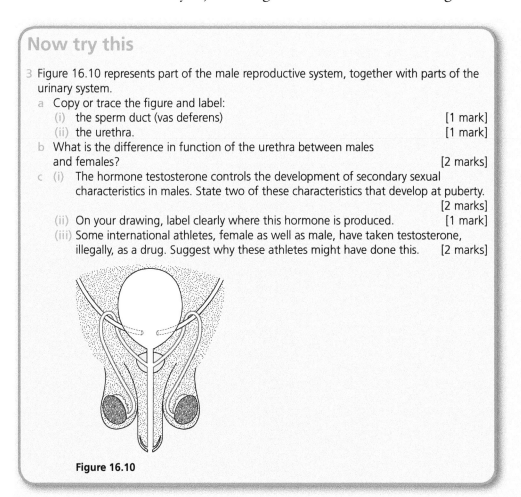

Figure 16.10

Production of oestrogen and progesterone

The sites of production of oestrogen and progesterone in the menstrual cycle and in pregnancy are listed in the following table.

Hormone	Site of production in the menstrual cycle	In pregnancy
Oestrogen	Ovaries	Placenta
Progesterone	Corpus luteum (remains of follicle in ovary after ovulation)	Placenta

Hormones and the menstrual cycle

The events in the menstrual cycle are shown in Figure 16.11.

At the start of the cycle, the lining of the uterus wall has broken down (menstruation). As each follicle in the ovaries develops, the amount of **oestrogen** produced by the ovary increases. The oestrogen acts on the uterus and causes its lining to become thicker and develop more blood vessels. These are changes that help an early embryo to implant. The **pituitary gland** at the base of the brain secretes **follicle-stimulating hormone (FSH)** and **luteinising hormone**, or **lutropin (LH)**, which promote ovulation.

Once the ovum has been released, the follicle that produced it develops into a solid body called the **corpus luteum**. This produces a hormone called **progesterone**, which makes the uterus lining grow thicker and produce more blood vessels. If the ovum is fertilised, the corpus luteum continues to release progesterone and so keeps the uterus in a state suitable for implantation. If the ovum is not fertilised, the corpus luteum stops producing progesterone. The lining of the uterus then breaks down and loses blood, which escapes through the cervix and vagina (menstruation).

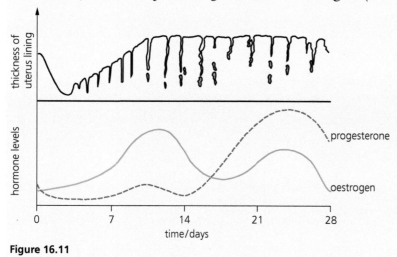

Figure 16.11

Methods of birth control

There are four main groups of birth control methods: natural, chemical, mechanical and surgical. Details are given in the following table.

Cambridge IGCSE Biology Study and Revision Guide Second Edition © Dave Hayward 2016

Type	Example	Details	Comments
Natural	Abstinence	No sexual intercourse	The best way of avoiding an unwanted pregnancy
	Rhythm method	The time of ovulation is predicted by monitoring body temperature and cervical mucus; intercourse is avoided around the date when these change	Time of ovulation can vary, so not always reliable
Chemical	Contraceptive pill	Contains progesterone and oestrogen, which prevent ovulation, or progesterone only (the 'mini-pill'), which prevents implantation of a blastula	Different strengths are available – a doctor needs to decide. Usually very reliable when taken regularly
	Intra-uterine device (IUD)	A plastic-coated copper coil is surgically inserted into the wall of the uterus	Prevents implantation of a blastula; reliable
	Intra-uterine system (IUS)	Similar to IUD, but releases progesterone	Prevents ovulation
	Implant	A small plastic tube containing progesterone is inserted under the skin of the upper arm	Prevents ovulation, last about three years; over 99% success rate
	Injection	Contains progesterone. Works by thickening the mucus in the cervix, acting as a barrier to sperm	Effective for 8–12 weeks
Mechanical	Condom	Rubber sheath placed over the penis to stop sperm entering the vagina	Also prevents transmission of sexually transmitted diseases; reliable if used with a spermicide
	Diaphragm	Dome-shaped rubber barrier that fits in the vagina at the cervix	Needs to be the correct size; must be left in place for six hours after intercourse; reliable if used with a spermicide
	Femidom	A thin plastic sheath placed inside the vagina	Also prevents transmission of sexually transmitted diseases; reliable if used with a spermicide
Surgical	Vasectomy	Sperm ducts are tied or cut so no sperm can leave the testes	Not normally reversible; extremely reliable
	Laparotomy (female sterilisation)	Oviducts are tied or cut so no eggs can pass down them	Not normally reversible; extremely reliable

Use of hormones for contraception and fertility treatments

Using hormones for contraception

Oestrogen and progesterone control important events in the menstrual cycle.

Oestrogen encourages the re-growth of the lining of the uterus wall after a period and prevents the release of FSH. If FSH is blocked, no further ova are matured. The uterus lining needs to be thick to allow successful implantation of an embryo. Progesterone maintains the thickness of the uterine lining. It also inhibits the secretion of LH, which is responsible for ovulation. If LH is suppressed, ovulation cannot happen, so there are no ova to be fertilised.

Because of the roles of oestrogen and progesterone, they are used, singly or in combination, in a range of contraceptive methods.

Social aspects of contraception and fertility treatments

Artificial insemination is a way of increasing the chances of a woman having a baby when the male partner is infertile. It involves using sperm from a donor. The sperm are inserted into the female partner's uterus around the time of ovulation.

Use of hormones in fertility drugs

Fertility drugs can be used to increase the chance of pregnancy. FSH and LH treatment causes multiple release of ova (eggs), increasing the chance of pregnancy.

Cambridge IGCSE Biology Study and Revision Guide Second Edition © Dave Hayward 2016

In vitro fertilisation is a method used if the woman has a problem with blocked oviducts. A doctor collects the ova produced by FSH and LH treatment. Some of the ova are fertilised in a Petri dish using the male partner's sperm. The early embryos produced are then inserted into the uterus to achieve pregnancy. The treatment is quite expensive, and not always successful.

Sexually transmitted infections

You need to learn the definition of a sexually transmitted infection (STI) that is given at the start of this chapter. These are diseases passed on during unprotected sexual intercourse. You need to know that **human immunodeficiency virus (HIV)** is an example of an STI.

HIV may result in acquired immune deficiency syndrome (AIDS). Details are shown in the following table.

Methods of transmission	Ways of preventing its spread
• Unprotected sexual intercourse with an infected person (this includes homosexuals) • Drug use involving sharing a needle used by an infected person • Transfusions of unscreened blood • From an infected mother to the fetus • Feeding a baby with milk from an infected mother • Use of unsterilised surgical instruments	• Use of a condom for sexual intercourse • Abstinence from sexual intercourse • Screening of blood used for transfusions • Use of sterilised needles for drug injections • Feeding a baby with bottled milk when the mother has HIV • Use of sterilised surgical instruments

How HIV affects the immune system

HIV attacks some types of lymphocyte (white blood cells) in the bloodstream. Lymphocytes produce antibodies, which attack the antigens present on invading microbes. Some lymphocytes are stored in lymph nodes to provide protection against future infections. HIV prevents this immunity being retained, so the AIDS sufferer has no protection against diseases such as tuberculosis (TB) and pneumonia.

Now try this

4 a HIV can be passed from mother to fetus through the placenta. State two other ways in which the virus can be passed to an uninfected person.   [2 marks]
   b Name two other harmful materials that might pass from mother to fetus through the placenta.   [2 marks]

Key objectives

The objectives for this chapter are to revise:

- definitions of the key terms
- mitosis and meiosis, including the role of mitosis
- that meiosis is involved in the production of gametes
- that a heterozygous individual cannot be pure-breeding
- how to interpret pedigree diagrams
- how to use genetic diagrams and Punnett squares to show monohybrid crosses
- the significance of the sequence of bases in a gene
- how DNA controls cell function and how a protein is made
- that all body cells in an organism contain the same genes, but many genes are not expressed
- the number of chromosomes in human haploid and diploid cells
- that the exact duplication of chromosomes occurs before mitosis and that, during mitosis, the copies of chromosomes separate
- that meiosis produces variation
- how to use a test-cross to identify an unknown genotype
- how to explain co-dominance, using the inheritance of ABO blood groups
- sex-linked inheritance, using colour blindness as an example
- how to use genetic diagrams to predict the results of monohybrid crosses involving co-dominance, sex-linkage and how to calculate phenotypic ratios

Key terms

Inheritance	The transmission of genetic information from generation to generation
Chromosome	A thread-like structure of DNA carrying genetic information in the form of genes
Gene	A length of DNA that codes for a protein
Allele	A version of a gene
Mitosis	Nuclear division that gives rise to genetically identical cells
Meiosis	Nuclear division that gives rise to cells that are genetically different
Genotype	The genetic make-up of an organism in terms of the alleles present
Phenotype	The observable features of an organism
Homozygous	Having two identical alleles of a particular gene
Heterozygous	Having two different alleles of a particular gene
Dominant	An allele that is expressed if it is present
Recessive	An allele that is expressed only when there is no dominant allele of the gene present
Haploid nucleus	A nucleus containing a single set of unpaired chromosomes, e.g. in gametes
Diploid nucleus	A nucleus containing two sets of chromosomes, e.g. in body cells
Meiosis	Reduction division in which the chromosome number is halved from diploid to haploid, resulting in genetically different cells
Sex-linked characteristic	A characteristic in which the gene responsible is located on a sex chromosome, which makes it more common in one sex than the other

Chromosomes, genes and proteins

Inheritance is the transmission of genetic information from one generation to the next, leading to continuity of the species and variation within it.

The definition of chromosomes is given at the start of this chapter. Figure 17.1 shows the relationship between a chromosome and the genes it carries.

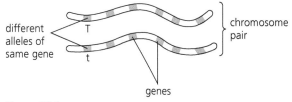

Figure 17.1

● Inheritance of sex in humans

Of the 23 pairs of chromosomes present in each human cell, one pair is the sex chromosomes. These determine the sex of the individual. Males have XY, females have XX. So, the presence of a Y chromosome results in male features developing. Figure 17.2 shows how sex is inherited.

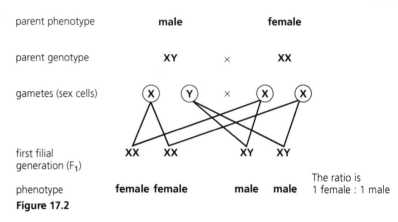

Figure 17.2

● The genetic code

The structure of DNA is described in Chapter 4. In summary:

- Each nucleotide carries one of four bases (A, T, C or G). A string of nucleotides therefore holds a sequence of bases. This sequence forms a code, which instructs the cell to make particular proteins. Proteins are made from amino acids linked together (Chapter 4).
- The type and sequence of the amino acids joined together will determine the kind of protein formed.
- Each group of three bases stands for one amino acid.
- A gene is a sequence of triplets of the four bases, which codes for one protein molecule.
- Insulin is a small protein with only 51 amino acids and so 153 (i.e. 3×51) bases in the DNA molecule. Most proteins are much larger than this and most genes contain a thousand or more bases.

The chemical reactions that take place in a cell determine the type of cell it is and what its functions are. These chemical reactions are, in turn, controlled by enzymes. Enzymes are proteins. So, by determining which proteins (particularly enzymes) are produced in a cell, the **genetic code** of DNA also determines the cell's structure and function. In this way, the genes also determine the structure and function of the whole organism. Other proteins coded for in DNA include antibodies and the receptors for neurotransmitters (see details of synapses in Chapter 14).

● The manufacture of proteins in cells

- DNA molecules remain in the nucleus, but the proteins that they carry the codes for are needed elsewhere in the cell.
- A molecule called messenger RNA (**mRNA**) is used to transfer the information from the nucleus.
- An RNA molecule is much smaller than a DNA molecule and is made up of only one strand.
- To pass on the protein code, the double helix of DNA unwinds to expose the chain of bases.

Cambridge IGCSE Biology Study and Revision Guide Second Edition © Dave Hayward 2016

- One strand acts as template. An mRNA molecule is formed along part of this strand, made up of a chain of nucleotides with complementary bases to a section of the DNA strand.
- The mRNA molecule carrying the protein code then moves out of the nucleus into the cytoplasm, where it passes through a **ribosome**.
- The mRNA molecule instructs the ribosome to put together a chain of amino acids in a specific sequence, thus making a protein. Other mRNA molecules will carry codes for different proteins.

Gene expression

Body cells do not all have the same requirements for proteins. For example, the function of some cells in the stomach is to make the protein pepsin (see 'Chemical digestion' in Chapter 7). Bone marrow cells make the protein haemoglobin, but do not need digestive enzymes. Specialised cells all contain the same genes in their nuclei, but only the genes needed to code for specific proteins are switched on (**expressed**). This enables the cell to make only the proteins it needs to fulfil its function.

Diploid nucleus and haploid nucleus

The definitions of diploid nucleus and haploid nucleus are given at the start of this chapter. In each diploid cell (nearly all body, or somatic, cells) there is a pair of each type of chromosome (see Figure 17.1). In a human diploid cell, there are 23 pairs. Sex cells (sperm and ova) are haploid, containing only 23 chromosomes. The 23 chromosomes comprise one from each pair. Each chromosome is made up of a large number of genes coding for the formation of different proteins that give us our characteristics.

Mitosis and meiosis

The definition of mitosis is given at the start of this chapter. Mitosis is a form of cell division used for making new cells to enable growth or the replacement of old or damaged cells. Asexual reproduction involves mitosis.

The definition of meiosis is given at the start of this chapter. Sex cells (gametes) are formed in the gonads (ovaries and testes) by meiosis. When ova are formed in a woman, all the ova will carry an X chromosome. When sperm are formed in a man, half the sperm will carry an X chromosome; half will carry a Y chromosome (see Figure 17.3).

Examiner's tip

Although many textbooks show the stages of mitosis and meiosis, you do not need those details for the Cambridge IGCSE Core or Extended exam.

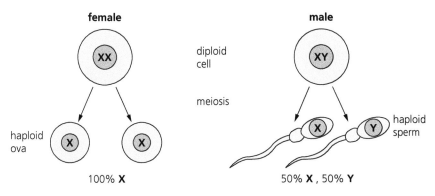

Figure 17.3 The formation of sex cells by meiosis

It is illegal to photocopy this page

● Sample question

Complete the following passage, using only words from the list below.

  diploid   gametes   haploid   meiosis   mitosis   red blood cells

The transfer of inherited characteristics to new cells and new individuals depends on two types of cell division. During _____, the chromosomes are duplicated exactly and _____ cells are produced.

  However, during _____, the chromosome sets are first duplicated and then halved, producing cells. These cells will become _____.

[4 marks]

Student's answer

During *meiotosis* ✖, the chromosomes are duplicated exactly and *identical* ✖ cells are produced.

However, during *meiosis* ✔, the chromosome sets are first duplicated and then halved, producing cells. These cells will become gametes. ✔

Examiner's comments

The first answer is not clear – it mixes up the terms 'mitosis' and 'meiosis'. Sometimes candidates do this deliberately when they are not sure of the answer, hoping that the examiner will give them the benefit of the doubt. (We do not!) This candidate has not followed the rubric (instructions) in the question for the second answer: the term 'identical' does not appear in the word list. The correct answers are 'mitosis' and 'diploid'.

● Mitosis

Before the process starts, all the chromosomes are duplicated exactly. During mitosis, the copies of the chromosomes separate and form two nuclei with the same number of chromosomes as the parent nucleus cell (the diploid number of chromosomes is maintained). At the end of a mitotic cell division, the number of cells is doubled and the daughter cells produced are genetically identical to the parent.

● Stem cells

Stem cells are unspecialised cells in the body that have retained their power of division by mitosis. The daughter cells produced can become specialised for specific functions. Examples include the basal cells of the skin, which keep dividing to make new skin cells, and cells in the red bone marrow, which constantly divide to produce the whole range of blood cells. Cells taken from early embryos (**embryonic stem cells**) can be induced to develop into almost any kind of cell. Scientists have learned to re-programme skin cells so that they develop into other types of cell, such as nerve cells.

● Meiosis

The definition of meiosis is given at the start of this chapter. Note that there is a more detailed version needed for the Extended paper.

Cambridge IGCSE Biology Study and Revision Guide Second Edition © Dave Hayward 2016

Meiosis is called a reduction division because it involves halving the normal chromosome number – the pairs of chromosomes are separated. The gametes (sex cells) produced are haploid, but they are formed from diploid cells. At the end of the process, the cells produced are not all identical – meiosis results in genetic variation.

Both the maternal and paternal chromosomes contain new combinations of genetic material.

Now try this

1 a The nuclei of human liver cells contain 46 chromosomes. Copy and complete the table below to show how many chromosomes would be present in the cells listed. [3 marks]

Type of cell	Number of chromosomes
Ciliated cell in windpipe	
Red blood cell	
Ovum	

b Describe two differences, other than the number of chromosomes, between nuclei produced by mitosis and those produced by meiosis. [2 marks]

● Monohybrid inheritance

You need to learn, and be able to use, definitions of six genetic terms: *genotype*, *phenotype*, *homozygous*, *heterozygous*, *dominant* and *recessive*. Definitions of these terms are given at the start of this chapter.

Monohybrid inheritance involves the study of how a single gene is passed on from parents to offspring. It is probably easiest to predict the outcome of a monohybrid cross using a Punnett square (see Figure 17.4). However, if you have been taught the traditional way of displaying the cross (as shown in Figure 17.2), there is nothing wrong with using that method.

All the genetic crosses shown will involve examples using pea plants, which can be tall (T) or dwarf (t) – tall is dominant to dwarf.

Examiner's tips

- When you write out a genetic cross, make sure you state what the symbols represent, e.g. T = tall, t = dwarf.
- Make sure you label each line in the cross (phenotype, genotype, etc.).
- It is a good idea to circle the gametes to show that meiosis has happened.
- Read the question really carefully – are you asked to state the outcome in terms of the genotype or the phenotype?

If two identical homozygous individuals are bred together, the product of the cross will be pure-breeding. However, if one parent is pure-breeding dominant and the other parent is pure-breeding dwarf, there will be a different outcome, as shown in the first example on the next page.

A cross between a pure-breeding tall pea plant and a pure-breeding dwarf pea plant:

As tall is dominant to dwarf, and both plants are pure-breeding, their genotypes must be TT and tt.

phenotypes of parents	**tall**		**dwarf**
genotypes of parents	**TT**	×	**tt**
gametes	(T) (T)	×	(t) (t)

Punnett square

	(T)	(T)
(t)	**Tt**	**Tt**
(t)	**Tt**	**Tt**

F₁ genotypes — all **Tt**

F₁ phenotypes — all tall

Figure 17.4

A cross between two heterozygous tall pea plants:

The genotype of both plants must be Tt.

phenotypes of parents	**tall**		**tall**
genotypes of parents	**Tt**	×	**Tt**
gametes	(T) (t)	×	(T) (t)

Punnett square

	(T)	(t)
(T)	**TT**	**Tt**
(t)	**Tt**	**tt**

F₁ genotypes — 1 **TT**, 2 **Tt** , 1 **tt**

F₁ phenotypes — **tall   tall   dwarf**

ratio — 3 tall : 1 dwarf

Figure 17.5

Note that, as shown above, a heterozygous individual will not be pure-breeding.

A cross between a heterozygous tall pea plant and a dwarf pea plant:

The heterozygous tall pea plant must be Tt.
The dwarf pea plant must be tt.

phenotypes of parents	**tall**		**dwarf**
genotypes of parents	**Tt**	×	**tt**
gametes	(T) (t)	×	(t) (t)

Punnett square

	(T)	(t)
(t)	**Tt**	**tt**
(t)	**Tt**	**tt**

F₁ genotypes — 2 **Tt**, 2 **tt**

F₁ phenotypes — **tall   dwarf**

ratio — 1 tall : 1 dwarf

Figure 17.6

● Common misconceptions

- Some students ignore the letters for alleles given in genetic questions and make up their own, without stating a key. This usually results in a number of marks being lost through errors that could easily have been avoided.

Cambridge IGCSE Biology Study and Revision Guide Second Edition © Dave Hayward 2016

2 In exam questions involving genetic crosses, you often need to predict the genotypes of the parents from descriptions of them. Work out the following genotypes, based on peas that can be round or wrinkled, with round being dominant to wrinkled. Remember that the dominant allele normally takes the capital letter of the characteristic it represents.
  a A heterozygous round pea. [1 mark]
  b A wrinkled pea. [1 mark]
  c A pure-breeding round pea. [1 mark]

● Pedigree diagrams and inheritance

The term pedigree often refers to the pure-breeding nature of animals, but is also used to describe human inheritance. Pedigree diagrams are similar to family trees and can be used to demonstrate how genetic diseases can be inherited. They can include symbols to indicate whether individuals are male or female and what their genotype is for a particular genetic characteristic.

One genetic disease is called cystic fibrosis. Cystic fibrosis sufferers tend to have a much shorter lifespan than normal and suffer from respiratory, digestive and reproductive problems.

A sufferer of cystic fibrosis has two recessive alleles (cc). A carrier of the disease has one normal allele and one recessive allele (Cc). A healthy non-carrier has two normal alleles (CC).

- A man who is not a carrier (CC) who has children with a woman who is not a carrier (CC) will produce 100% children who are not carriers (all CC).
- If one parent is a carrier for cystic fibrosis (Cc) and the other parent is not a carrier (CC), 50% of their children are likely to be carriers (Cc) and 50% will be not be carriers (CC).
- However, if both parents are carriers, then the likely ratio of offspring of non-carriers/carriers/cystic fibrosis sufferers (CC:Cc:cc) is 1:2:1. So, there is a 1 in 4 chance of a child born to these parents having cystic fibrosis.

The pedigree diagram (Figure 17.7) shows the inheritance of cystic fibrosis in a family.

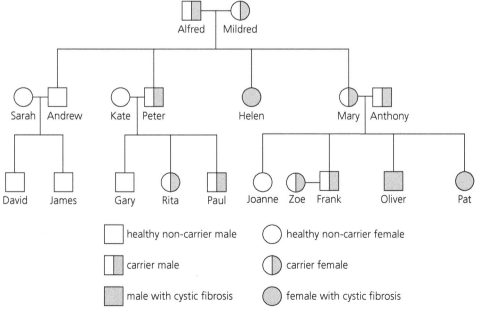

Figure 17.7 Pedigree diagram to show the inheritance of cystic fibrosis in a family

Parents Alfred and Mildred are married and both are cystic fibrosis carriers. However, because carriers have no symptoms of the disease, they may be unaware that they have defective alleles for cystic fibrosis. They go on to have four children. Three of these children eventually get married and have children of their own. One child, Helen, suffers from cystic fibrosis. The pedigree diagram shows that she does not get married and has no children.

Now try this

3 Copy and complete the passage by writing the most appropriate word from the list in each space.

chromosome   diploid   gene   heterozygous   meiosis
mutation   phenotype   recessive   dominant

Petal colour in pea plants is controlled by a single _____ that has two forms, red and white. The pollen grains are produced by _____. After pollination, fertilisation occurs and the gametes join to form a _____ zygote.

When two red-flowered pea plants were crossed with each other, some of the offspring had white flowers. The _____ of the rest of the offspring was red flowers. The white-flowered form is _____ to the red-flowered form and each of the parent plants was therefore _____.   [6 marks]

● Using a test-cross (back-cross)

A test-cross can be used to identify an unknown genotype. For example, a black mouse could have either the BB or the Bb genotype. One way to find out which it has is to cross the black mouse with a known homozygous recessive mouse (bb, having the phenotype of brown fur). The bb mouse will produce gametes with only the recessive b allele. A black homozygote (BB) will produce only B gametes.

BB × bb will produce 100% black individuals (all Bb).

Bb × bb will produce, on average, 50% black individuals (Bb) and 50% brown individuals (bb). This outcome identifies a parent that is not pure-breeding.

● Co-dominance

This term describes a pair of alleles, neither of which is dominant over the other. This means that both can have an effect on the phenotype when they are present together in the genotype. The result is that there can be three different phenotypes. When writing the genotypes of co-dominant alleles, the common convention is to use a capital letter to represent the gene involved, and a small raised (superscript) letter for each phenotype.

Example

The alleles of the gene for flower colour in a plant are C^R (red) and C^W (white). The capital letter C has been chosen to represent colour. Pure-breeding (homozygous) flowers may be red ($C^R C^R$) or white ($C^W C^W$). If these are cross-pollinated, all the first filial generation will be heterozygous ($C^R C^W$), which are pink because both alleles have an effect on the phenotype.

Self-pollinating the pink (F_1) plants results in an unusual ratio in the next (F_2) generation of red–pink–white of 1:2:1.

> **Now try this**
>
> 4 Write out a genetic cross for the example described above, to prove that the ratio achieved is 1:2:1. Use the same layout as in the examples given earlier. [6 marks]

● Common misconceptions

● When factors are co-dominant, students often think this will result in different proportions of offspring having the parents' features. However, co-dominance results in the appearance of a new characteristic, which is intermediate to the parents' features. For example, if the parents are pure-breeding for long fur and short fur, the offspring will all have medium-length fur.

● Inheritance of A, B, AB and O blood groups

These blood groups give an example of co-dominance. Instead of two alleles being present, in this case there are three: I^A, I^B and I^O. Combinations of these can result in four different phenotypes: A, B, AB and O. The alleles are responsible for producing antigens that respond to foreign antibodies (this can result in blood clotting in blood transfusions, and rejection of organs after transplant operations). However, while I^A and I^B are co-dominant, I^O is dominated by both the other alleles. This means, for example, that a person with blood group A could have the genotype $I^A I^A$ or $I^A I^O$. This has implications when having children because, if both parents carry the I^O allele, a child could be born with the genotype $I^O I^O$ (blood group O), even though neither of the parents have this phenotype.

Example: inheritance of blood group O

Two parents have blood groups A and B. The father is $I^A I^O$ and the mother is $I^B I^O$.

phenotypes of parents	**blood group A**	**blood group B**
genotypes of parents	$I^A I^O$ ×	$I^B I^O$
gametes	(I^A) (I^O) ×	(I^B) (I^O)

Punnett square

	(I^A)	(I^O)
(I^B)	$I^A I^B$	$I^B I^O$
(I^O)	$I^A I^O$	$I^O I^O$

F₁ genotypes	$I^A I^O$, $I^B I^O$, $I^A I^B$, $I^O I^O$
F₁ phenotypes	**A B AB O**
ratio	1 : 1 : 1 : 1

Figure 17.8

● Sex linkage

The definition of a sex-linked characteristic is given at the start of this chapter.

The sex chromosomes (X and Y) carry genes that control sexual development. In addition, they carry genes that control other characteristics. These tend to be on the X chromosome, which has longer arms to the chromatids. Even if the allele is recessive, because there is no corresponding allele on the Y chromosome, it is bound to be expressed in a male (XY). There is less chance of a recessive allele being expressed in a female (XX) because the other X chromosome may carry the dominant form of the allele.

One example of this is a form of colour blindness (Figure 17.9). In the following case, the mother is a carrier of colour blindness ($X^C X^c$). This means that she shows no symptoms of colour blindness, but the recessive allele causing colour blindness is present on one of her X chromosomes. The father has normal colour vision ($X^C Y$).

If the gene responsible for a particular condition is present on only the Y chromosome, only males can suffer from the condition because females do not possess the Y chromosome.

phenotypes of parents	mother: normal vision	father: normal vision

genotypes of parents		
	$X^C X^c$ ×	$X^C Y$

gametes

X^C X^c × X^C Y

Punnett square

$X^C X^c$

	X^C	X^c
X^C	$X^C X^C$	$X^C X^c$
Y	$X^C Y$	$X^c Y$

($X^C Y$ on the left)

F_1 genotypes: $X^C X^C$ $X^C X^c$ $X^C Y$ $X^c Y$

F_1 phenotypes: 2 females with normal vision; 2 males, one with normal vision, one with colour blindness

Figure 17.9

Cambridge IGCSE Biology Study and Revision Guide Second Edition © Dave Hayward 2016

Variation and selection

Key objectives

The objectives for this chapter are to revise:

- definitions of the key terms
- continuous and discontinuous variation
- the differences between phenotypic variation and genetic variation
- that mutation is the way in which new alleles are formed
- factors that can increase the rate of mutation
- the adaptive features of an organism from images of it
- natural selection and selective breeding
- the causes of phenotypic variation
- the symptoms of sickle cell anaemia
- the causes of the formation of abnormal haemoglobin and sickle-shaped red blood cells

- how to use genetic diagrams to show the inheritance of sickle cell anaemia
- how the presence of the sickle cell allele can provide resistance to malaria
- the distribution of the sickle cell allele in relation to the distribution of malaria
- how to explain the adaptive features of hydrophytes and xerophytes
- evolution
- the development of strains of antibiotic-resistant bacteria
- the difference between natural and artificial selection
- how selective breeding by artificial selection is used to improve crop plants and domesticated animals

● Key terms

Variation	The differences between individuals of the same species
Mutation	Genetic change
Adaptive feature	An inherited feature that helps an organism to survive and reproduce in its environment
Gene mutation	A change in the base sequence of DNA
Adaptive feature	The inherited functional features of an organism that increase its fitness
Fitness	The probability of an organism surviving and reproducing in the environment in which it is found
Process of adaptation	The process, resulting from natural selection, by which populations become more suited to their environment over many generations

● Variation

The term *variation* is defined at the start of this chapter. Those variations that can be inherited are determined by genes. They are **genetic variations**. **Phenotypic variations** may be brought about by genes, but can also be caused by the environment, or a combination of both genes and the environment. There are two main types of variation: continuous and discontinuous.

Continuous variation

Continuous variation shows a complete range of the characteristic within a population between two extremes. It is caused both by genes (often a number of different genes) and by the environment. Environmental influences for plants may be the availability of, or competition for, nutrients, light and water and exposure to disease. For animals, environmental influences can include the availability of food or balanced diet, exposure to disease (or the availability of health services for humans), etc.

Examples of continuous variation include height, body mass and intelligence. When the frequency is plotted on a graph as in Figure 18.1, a smooth curve is produced, with the majority of the population sample grouped together and only small numbers at the extremes of the graph.

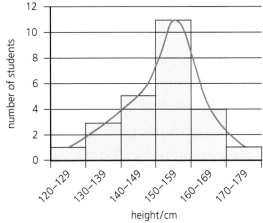

Figure 18.1

Discontinuous variation

Discontinuous variation is seen where there is a limited number of obvious, distinct categories for a feature. There are no intermediates between categories, and the feature cannot usually change during life. It is caused by a single gene or a small number of genes, with no environmental influence (as in Figure 18.2).

Examples include blood group, ability to tongue-roll and earlobe shape. When the frequencies are plotted on a graph, bars are produced that cannot be linked with a smooth curve.

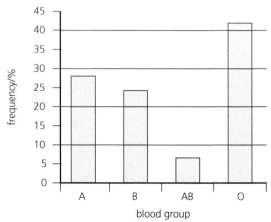

Figure 18.2

● Common misconceptions

● When plotting graphs of discontinuous variation, students often place the bars in contact with each other. These graphs should be drawn with a gap between each bar because they represent distinct categories (see Figure 18.2). This is called a bar graph. However, on a graph of continuous variation, there is one characteristic being plotted with a range of numerical values, so the bars do touch (see Figure 18.1). This is called a histogram.

● Sample question

Seventy seeds were collected from a cross between two plants of the same species. The seeds were sown at the same time and, after three weeks, the heights of the plants that grew were measured and found to fall into two groups, A and B, as shown in Figure 18.3.

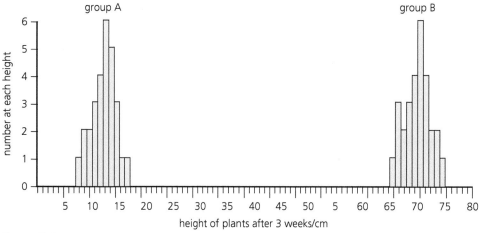

Figure 18.3

1 Calculate the percentage of seeds that germinated. Show your working.   [2 marks]

2 a Name the type of variation shown within each group.   [1 mark]

  b State three factors that might have caused this variation.   [3 marks]

Student's answer

1 *56 / 70 × 100 = 80%* ✓✓
2 *a continuous variation* ✓
   b disease ✓*, temperature* ✓*, amount of light, species of plant* ✗

Examiner's comments

The answers for 1 and 2a were good. However, there were three marks for 2b and the candidate gave four answers. As the last answer was incorrect, the third mark was not received.

● Mutation

Mutation is a source of variation, caused by an unpredictable change in the genes or chromosome numbers. As a result, new alleles are formed.

Mutations are normally very rare. However, exposure to radiation and some chemicals, such as tar in tobacco smoke, increases the rate of mutation. Exposure can cause uncontrolled cell division, leading to the formation of tumours (cancer). The exposure of gonads (testes and ovaries) to radiation can lead to sterility or damage to genes in sex cells that can be passed on to children.

● Variation

Phenotypic variation is caused by both genetic and environmental factors.

For example, a fair-skinned person may be able to change the colour of his or her skin by exposing it to the Sun, so getting a tan. The tan is an **acquired characteristic**.

You cannot inherit a suntan. Black skin, on the other hand, is an **inherited characteristic**. Many features in plants and animals are a mixture of acquired and inherited characteristics. For example, some fair-skinned people never go brown in the Sun, they only become sunburnt. They have not inherited the genes for producing the extra brown pigment in their skin. A fair-skinned person with the genes for producing pigment will only go brown if he or she exposes themselves to sunlight. So, the tan is a result of both inherited and acquired characteristics.

Discontinuous variation is mostly caused by genes alone. An example is blood groups in humans (Figure 18.2). Environmental factors will not cause a change in a person's blood group.

A gene mutation (see definition at the start of this chapter) can result in a genetic change. The sequence of bases in DNA becomes altered, resulting in a change in coding for one or more amino acids (see Chapter 4). A section of DNA may now start making a different protein that could affect the organism.

● Sickle cell anaemia

Sickle cell anaemia is caused by a mutation in the blood pigment haemoglobin. The defective haemoglobin molecule differs from normal haemoglobin by only one amino acid (represented by a sequence of three bases). When the faulty haemoglobin is present in a red blood cell, it causes the cell to deform and become sickle shaped, especially when oxygen levels in the blood become low. In this state, the sickled red blood cells are less efficient at transporting oxygen and are more likely to become stuck in a capillary, preventing blood flow. The distortion and destruction of the red cells, which occurs in low oxygen concentrations, leads to periods of severe anaemia.

The faulty allele (Hb^S) is dominated by the allele for normal haemoglobin, but still has some effect in a heterozygous genotype. The possible genotypes are:

- Hb^AHb^A – normal haemoglobin, no anaemia;
- Hb^AHb^S – some abnormal haemoglobin, sickle cell trait (not life-threatening);
- Hb^SHb^S – abnormal haemoglobin, sickle cell anaemia (life-threatening).

Genetic diagrams can be used to show how sickle cell anaemia is inherited in the same way as other genetic characteristics.

For example, the cross between two parents who are both heterozygous for sickle cell anaemia is shown in Figure 18.4.

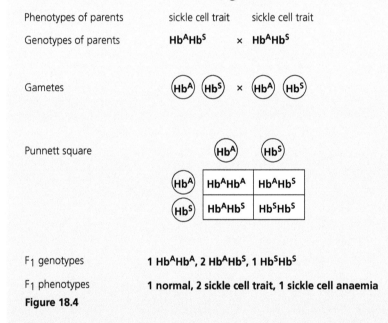

Phenotypes of parents   sickle cell trait   sickle cell trait

Genotypes of parents   **Hb^AHb^S**   ×   **Hb^AHb^S**

Gametes Hb^A Hb^S × Hb^A Hb^S

Punnett square

	Hb^A	Hb^S
Hb^A	Hb^AHb^A	Hb^AHb^S
Hb^S	Hb^AHb^S	Hb^SHb^S

F₁ genotypes   **1 Hb^AHb^A, 2 Hb^AHb^S, 1 Hb^SHb^S**

F₁ phenotypes   **1 normal, 2 sickle cell trait, 1 sickle cell anaemia**

Figure 18.4

Malaria is a life-threatening disease caused by a parasite that invades red blood cells. The parasite is carried by some species of mosquito. However, a person who is heterozygous (Hb^AHb^S) for sickle cell anaemia has protection from malaria, because the malaria parasite is unable to invade and reproduce in the sickle cells. A person who is homozygous for sickle cell anaemia (Hb^SHb^S) also has protection, but is at a high risk of dying from sickle cell anaemia. A person with normal haemoglobin (Hb^AHb^A) in a malarial country is at a high risk of contracting malaria.

When the distributions of malaria and sickle cell anaemia are shown on a map of the world, it is found that the two coincide in tropical areas because of the selective advantage of the Hb^S allele in providing protection against malaria.

● Adaptive features

The term *adaptive feature* is defined at the start of this chapter. You need to be able to interpret images and other information about a species to describe its adaptive features.

Cambridge IGCSE Biology Study and Revision Guide Second Edition © Dave Hayward 2016

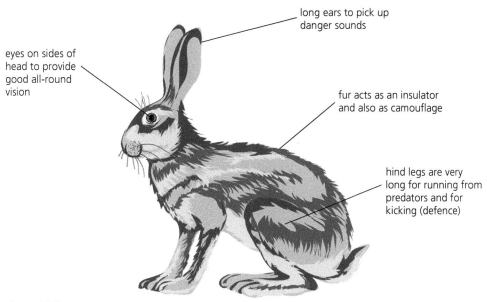

eyes on sides of head to provide good all-round vision

long ears to pick up danger sounds

fur acts as an insulator and also as camouflage

hind legs are very long for running from predators and for kicking (defence)

Figure 18.5

Examiner's tip

- Print or copy a photograph or diagram of an organism (animal or plant) and annotate the illustration to identify and describe the organism's adaptive features. An example (hare) is shown in Figure 18.5.
- In your description, always state the adaptation and how it helps the organism to survive.

● Adaptive features and fitness

Definitions of the terms *adaptive feature* and *fitness* are given at the start of this chapter. Note that there is a slightly different definition of *adaptive feature* in the Core and Extended syllabus.

● How hydrophytes and xerophytes are adapted to their environments

Where possible, you should be able to describe these features based on plants you are familiar with and that grow in your local area. Some plants are adapted to cope with a lack of water (e.g. in very dry or 'arid' environments). These are called **xerophytes**.

Plant	Modifications
Pinus (pine tree)	Leaves are needle-shaped to reduce surface area for transpiration and to resist wind damage
	Sunken stomata to create high humidity and reduce transpiration
	Thick waxy cuticle on the epidermis to prevent evaporation from leaf surface

Some plants are adapted to cope with living in water (e.g. pond plants and seaweeds). These are called **hydrophytes**.

Plant	Modifications
Nymphaea (water lily)	Roots are poorly developed and contain air spaces (owing to poor oxygen levels in the mud they grow in)
	Leaves contain large air spaces to make them buoyant, so they float on or near the surface (to gain light for photosynthesis)
	Lower epidermis of the leaves lack stomata to avoid waterlogging, while the upper epidermis has stomata for gas exchange
	Stems lack much support as the water surrounding them provides buoyancy for the plant

Selection

Natural selection

Variation describes differences in a population. Some variation is inherited (passed on from parents) and some is acquired (developed during life).

Animals and plants produced by sexual reproduction will show variation from their parents, for example in the size of the muscles in the legs of lions.

When organisms reproduce, many offspring are often produced. However, not all of them are likely to survive because of competition for resources such as food, water and shelter. The same is true for plants (they compete for resources such as nutrients, light, water and space). There is a struggle for survival.

The individuals with the most favourable characteristics are most likely to survive because they have an advantage over others in the population. For example, a lion cub with bigger muscles in its legs would be able to run more quickly and get food more successfully than its siblings.

In an environment where there is a food shortage, the individual with the best adaptations to the environment is most likely to survive to adulthood. The weaker individuals die before having the chance to breed, but the surviving adults breed and pass on the advantageous alleles to their offspring. More of the next generation carry the advantageous genes, resulting in a stronger population, better adapted to a changing environment.

Selective breeding

Selective breeding is used by humans to produce varieties of animals and plants that have an increased economic importance:

- Humans first select individuals with desirable features.
- These individuals are cross-bred to produce the next generation.
- From that generation, the offspring with desirable features are selected for further breeding.

Evolution

Slow changes in the environment result in adaptations in a population to cope with the change. Failure to adapt could result in the species becoming extinct. This gradual change in the species through **natural selection** over time, in response to changes in the environment, is a possible mechanism for **evolution**.

The definition of **the process of adaptation** is given at the start of this chapter.

The development of strains of antibiotic-resistant bacteria

This is an example of evolution by natural selection.

Bacteria reproduce rapidly – a new generation can be produced every 20 minutes by binary fission (see Chapter 16). Antibiotics are used to treat bacterial infections. An antibiotic is a chemical that kills bacteria by preventing bacterial cell wall formation. Mutations occur during reproduction, which produce some variation in the population of bacteria. Individual bacteria with the most favourable features are most likely to survive send reproduce. A mutation may occur that enables a bacterium

Cambridge IGCSE Biology Study and Revision Guide Second Edition © Dave Hayward 2016

toresist being killed by antibiotic treatment, while the rest of the population is killed when treated. This bacterium would survive the treatment and breed, passing on the antibiotic resistance gene to its offspring. Future treatment of this population of bacteria using the antibiotic would be ineffective.

Differences between natural and artificial selection

Natural selection occurs in groups of living organisms through the passing on of genes to the next generation by the best-adapted organisms, without human interference. Those with genes that provide an advantage to cope with changes in environmental conditions are more likely to survive, while others die before they can breed and pass on their genes. However, variation within the population remains.

Selective breeding is used by humans to produce varieties of animals and plants that have an increased economic importance. It is considered a safe way of developing new strains of organisms and is a much faster process than natural selection. However, selective breeding removes variation from a population, leaving it susceptible to disease and unable to cope with changes in environmental conditions. Potentially, therefore, selective breeding puts a species at risk of extinction.

Examples of improving crop plants and domesticated animals by selective breeding

Wild varieties of plants sometimes have increased resistance to fungal diseases, but have poor fruit yield. Cross-breeding wheat plants over a number of generations and selecting the organisms with the best features at each stage can result in the formation of varieties that have both high resistance to disease and high seed yield.

A variety of cattle may have a higher than average milk yield. Another variety may have a very high meat yield. If the two varieties are cross-bred, the individuals in the next generation with the best features are selected to continue breeding until a new breed has been artificially produced with the benefits of both parental varieties (high milk production in females; high meat yield in males).

Now try this

1 Farmers have carried out selective breeding to improve the breeds of some animals. Some of the original breeds have become very rare and are in danger of becoming extinct.
   a Explain the meaning of selective breeding. [2 marks]
   b Outline how selective breeding has been used to develop a named variety of animal or plant. State the characteristics of the new variety. [4 marks]

Key objectives

The objectives for this chapter are to revise:

- definitions of the key terms
- the role of the Sun in biological systems
- that energy is transferred between organisms in a food chain by ingestion
- how to construct simple food chains and how to interpret them
- how to use food chains and food webs to describe the impacts humans have
- how to draw, describe and interpret pyramids of numbers
- the carbon and water cycles
- the effects of burning fossil fuels and cutting down forests
- the factors affecting the rate of population growth for a population of an organism
- the increase in human population over the past 250 years
- how to interpret graphs and diagrams of human population growth
- the flow of energy through living organisms and how energy is transferred between trophic levels and why this transfer is inefficient
- why food chains usually have fewer than five trophic levels
- how to identify and name the trophic levels in food webs, food chains, pyramids of numbers and pyramids of biomass
- how to describe and interpret pyramids of biomass
- the advantages and disadvantages of using a pyramid of biomass rather than a pyramid of numbers to represent a food chain
- the nitrogen cycle and the roles of microorganisms in it
- how to identify the phases of a sigmoid population growth curve and the factors that lead to each phase

● Key terms

Food chain	The transfer of energy from one organism to the next, beginning with the producer
Food web	A network of interconnected food chains
Producer	An organism that makes its organic nutrients, usually using energy from sunlight through photosynthesis
Consumer	An organism that gets its energy by feeding on other organisms
Herbivore	An animal that gets its energy by eating plants
Carnivore	An animal that gets its energy by eating other animals
Decomposer	An organism that gets its energy from dead or waste material
Population	A group of organisms of one species, living in the same area, at the same time
Trophic level	The position of an organism in a food chain, food web, pyramid of numbers or pyramid of biomass
Community	All of the populations of different species in an ecosystem
Ecosystem	A unit containing a community of organisms and their environment, interacting together, e.g. a decomposing log or a lake

Now try this

1 State the difference between the terms in each of the following pairs:
  a *producer* and *consumer*   [4 marks]
  b *carnivore* and *herbivore*.   [2 marks]

● Energy flow

The Sun is the principal source of energy input to biological systems. The Earth receives two main types of energy from the Sun: light (solar) and heat. Photosynthetic plants and some bacteria can trap light energy and convert it into chemical energy.

You need to be able to define the first eight terms in the 'Key terms' of this chapter.

Heterotrophic organisms obtain their energy by eating plants or animals that have eaten plants. Thus, all organisms, directly or indirectly, get their energy from the Sun. The chemical energy produced is passed from one organism to another in a food chain but, unlike water and elements such as carbon and nitrogen, energy does not return in a cycle. The energy given out by organisms is lost to the environment.

● Food chains

Energy is transferred between organisms in a food chain by **ingestion**. **Food chains** are lists of organisms that show the feeding relationship between them, as in the example below.

| maize | ⟶ | locust | ⟶ | lizard | ⟶ | snake |
| *producer* | | *primary consumer* | | *secondary consumer* | | *tertiary consumer* |

Examiner's tip

When writing out a food chain, don't include the Sun (it is not an organism).

A food chain usually starts with a producer (photosynthetic plant), which gains its energy from the Sun. The arrows used to link each organism to the next represent the transfer of energy. They always point towards the 'eater' and away from the plant. The feeding level is known as the **trophic level**.

- Plants are producers (they make, or produce, food for other organisms).
- Animals that eat plants are primary consumers (a consumer is an 'eater'). They are also called herbivores.
- Animals that eat other animals are secondary, or possibly tertiary, consumers, depending on their position in the chain. They are also called carnivores.

Examiner's tips

- Make sure you can write a food chain involving three consumers, with the arrows in the correct direction.
- Always start with the producer on the left of the diagram.
- Practise labelling each trophic level in your food chain under the organisms (producer, primary consumer, etc.).
- Do not waste time drawing the plants and animals: this will not get you any extra marks.

● Food webs

Food webs are a more accurate way of showing feeding relationships than food chains, because most animals have more than one food source. For example, in the food web in Figure 19.1, the leopard feeds on baboons and impala.

The producer is grass. Locusts and impala are primary consumers (herbivores) and the baboon is a tertiary consumer (carnivore). The leopard is acting as a secondary and a fourth consumer.

Food chains and webs are easily

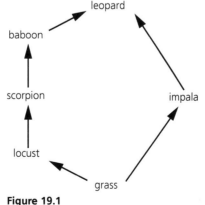

Figure 19.1

unbalanced, especially if one population of organisms in the web dies or disappears. Humans can be the cause, through over-harvesting (perhaps by over-predation or hunting) or through the introduction of foreign species to a habitat.

For example, in the food web in Figure 19.1, if all the baboons were killed by hunters, the leopard would have only impala to eat and so the impala population would decrease. The scorpion population may increase because of less predation by baboons, but if there are more scorpions they will eat more locusts, reducing the locust population, and so on.

● Common misconceptions

● Drawing food chains and webs with the arrows the wrong way round or putting the chain back-to-front (or both). Food chains and webs should start with the producer. The following example was seen in a recent paper:

jackal ⟶ sheep ⟶ grass

This student is suggesting that grass eats sheep and sheep eat jackals!

Now try this

2 Figure 19.2 shows a food web.

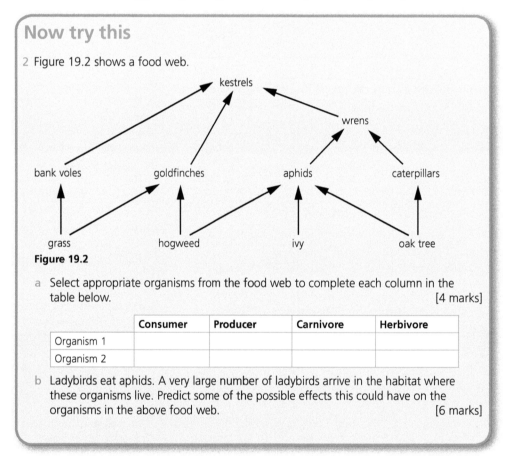

Figure 19.2

a Select appropriate organisms from the food web to complete each column in the table below. [4 marks]

	Consumer	Producer	Carnivore	Herbivore
Organism 1				
Organism 2				

b Ladybirds eat aphids. A very large number of ladybirds arrive in the habitat where these organisms live. Predict some of the possible effects this could have on the organisms in the above food web. [6 marks]

● Food pyramids

In a food pyramid, each trophic level in a food chain is represented by a horizontal bar, with the width of the bar representing the number of organisms at that level. The base of the pyramid represents the producer; the second level is the primary consumer; and so on.

● Pyramids of numbers

Figure 19.3 shows a typical pyramid of numbers.

Usually, the producers have the largest numbers, so they form the widest bar. There will be fewer primary consumers, and even fewer secondary consumers, so a pyramid shape is formed. However, this is not always true. Figure 19.4 shows a different pyramid of numbers.

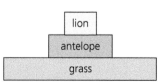

Figure 19.3

Cambridge IGCSE Biology Study and Revision Guide Second Edition © Dave Hayward 2016

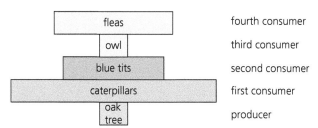

Figure 19.4

The food chain of Figure 19.4 is supported by a single organism (a large oak tree). Many caterpillars feed on its leaves. Only a single owl is supported by the blue tits. However, the owl has many fleas, which feed on it by sucking its blood.

● Food chains and energy

Energy is lost at each level in the food chain, as it is transferred between trophic levels.

The following examples show how the energy is lost:

- Energy lost through the process of respiration (as heat).
- Energy used up for movement (to search for food, find a mate, escape from predators, etc.).
- Warm-blooded animals (birds and mammals) maintain a constant body temperature – they lose heat to the environment.
- Warm-blooded animals lose heat energy in faeces and urine.
- Some of the material in the organism being eaten is not used by the consumer; for example, a locust does not eat the roots of maize, and some of the parts eaten are not digestible.

Even plants do not make use of all the light energy available to them. This is because some light:

- is reflected off shiny leaves;
- is the wrong wavelength for chlorophyll to trap;
- passes through the leaves without passing through any chloroplasts;
- does not fall on the leaves.

This means that the transfer of energy between trophic levels is inefficient – a lot is lost. On average, about 90% of the energy is lost at each level in a food chain. This means that, in long food chains, very little of the energy entering the chain through the producer is available to the top carnivore. Thus, there tend to be small numbers of top carnivores, and food chains usually have fewer than five trophic levels.

The food chain below shows how energy reduces through the chain. It is based on maize obtaining 100 units of energy.

maize ⟶ locust ⟶ lizard ⟶ snake
100 units   10 units   1 unit   0.1 unit

In shorter food chains, less energy is lost. In terms of conservation of energy, short food chains are more efficient than long ones in providing energy to the top consumer. Below are two food chains and the energy values for each level in them. Both food chains have a human being as the top consumer.

maize ⟶ cow ⟶ human
100 units   10 units   1 unit

maize ⟶ human
100 units   10 units

Ten times more energy is available to the human in the second food chain than in the first. In the second food chain, the human is a herbivore (vegetarian).

Some farmers try to maximise meat production by reducing movement of their animals (keeping them in pens or cages with a food supply) and keeping them warm in winter. This means less stored energy is wasted by the animals.

Examiner's tip

When describing food chains, food webs and pyramids, try to use the 'Key terms' defined at the start of this chapter wherever possible. Answers containing correct terminology used appropriately are more likely to earn you high marks.

Now try this

3 Figure 19.5 shows the flow of energy through a complete food chain.

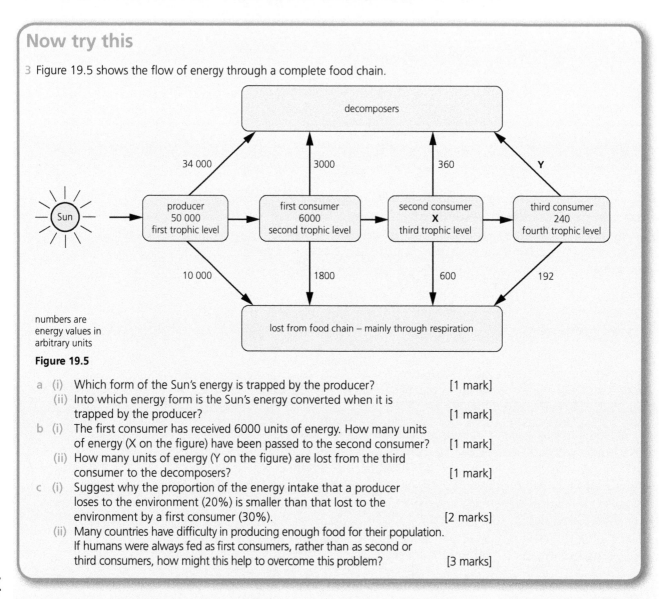

numbers are energy values in arbitrary units

Figure 19.5

a (i) Which form of the Sun's energy is trapped by the producer? [1 mark]
   (ii) Into which energy form is the Sun's energy converted when it is trapped by the producer? [1 mark]
b (i) The first consumer has received 6000 units of energy. How many units of energy (X on the figure) have been passed to the second consumer? [1 mark]
   (ii) How many units of energy (Y on the figure) are lost from the third consumer to the decomposers? [1 mark]
c (i) Suggest why the proportion of the energy intake that a producer loses to the environment (20%) is smaller than that lost to the environment by a first consumer (30%). [2 marks]
   (ii) Many countries have difficulty in producing enough food for their population. If humans were always fed as first consumers, rather than as second or third consumers, how might this help to overcome this problem? [3 marks]

● Pyramids of biomass

Figure 19.4 shows that pyramids of numbers are limited in what they show. It is more useful to measure the amount of living material (biomass) at each level over a fixed area of habitat. Once this is done, a normal-shaped pyramid is usually obtained, as shown in Figure 19.6.

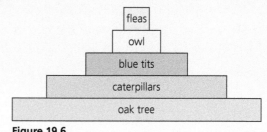

Figure 19.6

You need to be able to identify the trophic levels in food chains, food webs, pyramids of numbers and pyramids of biomass. Sometimes (as in Figure 19.1) there is a fourth level of consumer. This is called a **quaternary consumer**.

● Nutrient cycles

The carbon cycle

Figure 19.7 shows the main parts of the carbon cycle.

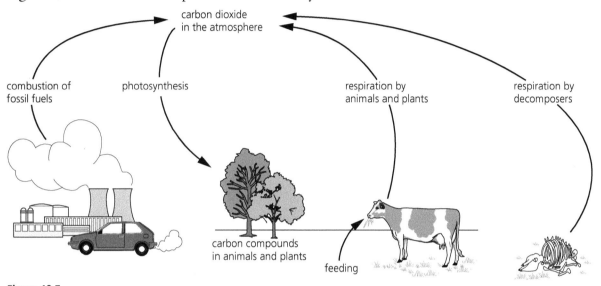

Figure 19.7

- Carbon moves into and out of the atmosphere mainly in the form of carbon dioxide.
- Plants take carbon dioxide out of the air by photosynthesis.
- Plants convert carbon dioxide into organic materials (carbohydrates, fats and proteins).
- Herbivores obtain carbon compounds by feeding on plants. Carnivores gain carbon compounds by feeding on other animals.
- Animals and plants release carbon dioxide back into the air through respiration.
- When organisms die, they usually rot (the process of decomposition). Decomposers break down the organic molecules through the process of respiration to release energy. This also releases carbon dioxide into the air.
- If a dead organism does not decompose, the carbon compounds are trapped in its body. Over a long period, this can form fossil fuels such as coal, oil or gas (fossilisation).
- Combustion of fossil fuels releases carbon dioxide back into the air.

Examiner's tips

- Do not be put off by complicated diagrams of the carbon cycle. Five main processes are involved: photosynthesis, respiration, decomposition, fossilisation and combustion.
- Study the word equations for respiration and photosynthesis: they are basically the same, but reversed.

Now try this

4 Figure 19.8 shows a diagram of the carbon cycle.

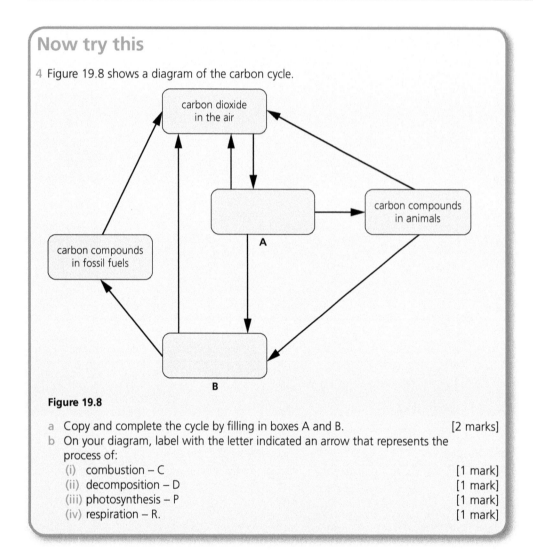

Figure 19.8

a Copy and complete the cycle by filling in boxes A and B. [2 marks]
b On your diagram, label with the letter indicated an arrow that represents the process of:
  (i) combustion – C [1 mark]
  (ii) decomposition – D [1 mark]
  (iii) photosynthesis – P [1 mark]
  (iv) respiration – R. [1 mark]

● Common misconceptions

● Plants do not start respiring when they stop photosynthesising (at night) – they respire all the time, but during the day there is usually a net intake of carbon dioxide and output of oxygen.

If there is an increase in the combustion of fossil fuels or if more trees are cut down and not replaced, carbon dioxide levels in the atmosphere will increase. This is thought to contribute to global warming. Carbon dioxide forms a layer in the atmosphere, which traps heat radiation from the Sun. This causes a gradual increase in the atmospheric temperature which can:

● melt polar ice caps, causing flooding of low-lying land;
● change weather conditions in some countries, increasing flooding or reducing rainfall and changing arable (farm) land to desert;
● cause the extinction of some species that cannot survive at higher temperatures.

The water cycle

Figure 19.9 shows the main parts of the water cycle.

● Plants release water vapour into the air through transpiration.
● Water evaporates from seas, lakes, rivers and soil.
● Water vapour condenses in the air, forming clouds.

- Water returns to the land as rain (precipitation), draining into streams, rivers, lakes and seas.
- Plant roots take up water by osmosis.

In addition, animals lose water to the environment through exhaling and sweating, and in urine and faeces.

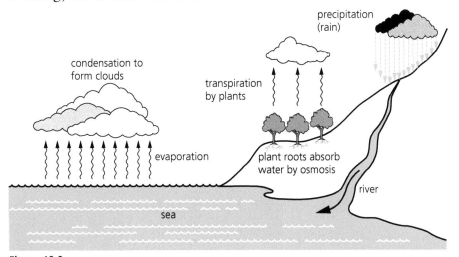

Figure 19.9

The nitrogen cycle

Figure 19.10 shows the main parts of the nitrogen cycle. You do not need to know the names of individual bacteria, but you do need to know the roles of the three main types:

- Nitrogen-fixing bacteria – convert nitrogen gas into compounds of ammonia.
- Nitrifying bacteria – convert compounds of ammonia into nitrates.
- Denitrifying bacteria – break down nitrites into nitrogen gas.

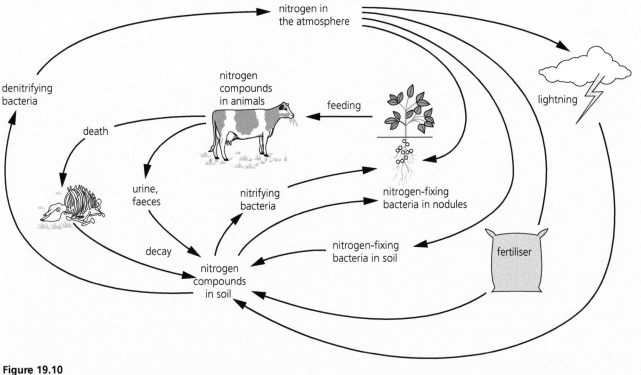

Figure 19.10

125

The element nitrogen is a very unreactive gas. Plants are not able to change it into nitrogen compounds, but it is needed to form proteins. Nitrogen compounds become available for plants in the soil in a number of ways, including:

- nitrogen-fixing bacteria (some plants – legumes such as peas, beans and clover – have roots with nodules that contain these bacteria, so the plant receives a direct source of nitrates);
- breakdown of dead plants and animals by decomposers (bacteria, fungi and invertebrates);
- the addition of artificial fertilisers, compost (decaying plant material) and manure (decaying animal waste – urine and faeces);
- lightning – its energy causes nitrogen to react with oxygen.

Plants absorb nitrates into their roots by active uptake (see Chapter 3). The nitrates are combined with glucose (from photosynthesis) to form amino acids and proteins. Proteins are passed through the food chain as animals eat the plants. When animals digest proteins, the amino acids released can be reorganised to form different proteins.

Some soil bacteria – denitrifying bacteria – break down nitrogen compounds and release nitrogen back into the atmosphere. This is a destructive process, commonly occurring in waterlogged soil. Farmers try to keep soil well drained to prevent this happening – a shortage of nitrates in the soil stunts the growth of crop plants.

Nitrates and other ammonium compounds are very soluble, so they are easily leached out of the soil and can cause pollution (see Chapter 21).

● Sample question

Figure 19.11 shows the nitrogen cycle.

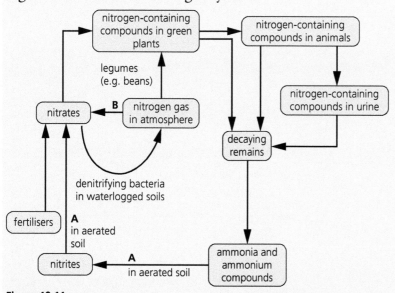

Figure 19.11

1 a Name the main nitrogen-containing compound found both in plants and in animals. [1 mark]

  b Name one nitrogen-containing compound that is present in urine. [1 mark]

  c Name the type of organism that causes the changes at A. [1 mark]

  d What atmospheric conditions bring about the change at B? [1 mark]

Cambridge IGCSE Biology Study and Revision Guide Second Edition © Dave Hayward 2016

2 Using the figure, explain why it is an advantage to have good
drainage in most agricultural land. [4 marks]

Student's answer

1 a Protein ✓
  b Urea ✓
  c Microbes ✗
  d Lightning ✓
2 Waterlogged soils contain denitrifying bacteria, ✓ which change nitrates to
  nitrogen gas. ✓ Most plants cannot use nitrogen and they would be short of
  nitrates for them to absorb. ✓

Examiner's comments

The answers to part 1 have gained three out of the four marks available. Microbes is too
general for 1c – bacteria was the response needed. The response to part 2 is good, but
a fourth mark was available for stating what plants use the nitrate for (formation of
proteins) or the effect of a shortage of nitrates on the plant (poor growth).

● Population size

Factors affecting the rate of population growth

The rate of growth of a population depends on the following.

Food supply – ample food will enable organisms to breed more
successfully to produce more offspring; a shortage of food can result in death
or can force emigration, reducing the population.

Predation – if there is heavy predation of a population, the breeding rate
may not be sufficient to produce enough organisms to replace those eaten, so
the population will drop in numbers. There tends to be a time lag in
population size change for predators and their prey. As predator numbers
increase, prey numbers drop, and as predator numbers drop, prey numbers
rise again (unless there are other limiting factors).

Disease – this is a particular problem in large populations, because disease
can spread easily from one individual to another. Epidemics can reduce
population sizes very rapidly.

Human population growth

Human population growth has been the result of having no limiting
factors. In 2014, the human population size was 7.2 billion. Human
population size has increased exponentially (Figure 19.12) because
of improvements in food supply and the development of medicine to
control diseases. Infant mortality has decreased, while life expectancy
has increased. Such a rapid increase in population size has social
implications. These include increasing demands for basic resources
including food, water, space, medical care and fossil fuels. The
presence of a larger human population creates greater pressures on the
environment (more land needed for housing, growing crops and road
building, as well as wood for fuel and housing) and, potentially, more
pollution. The presence of a larger population of young people results in
greater demands on education, while more old people results in greater
demands on health care.

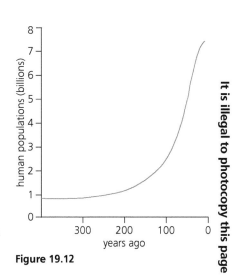

Figure 19.12

Examiner's tips

You need to be able to interpret graphs like that in Figure 19.12. This may involve describing the pattern shown, or quoting data extracted from the graph. Always try to support a description with data.

You need to be able to define the terms *community* and *ecosystem*.

The terms population, community and ecosystem are associated, as shown below.

individuals of the same species } = POPULATION + populations of other species } = COMMUNITY non-living part of environment + = COMMUNITY } = ECOSYSTEM

In a lake, the animal **community** will include **populations** of fish, insects, crustaceans, molluscs and protoctists. The plant community will consist of rooted plants with submerged leaves, rooted plants with floating leaves, reed-like plants growing at the lake margin, plants floating freely on the surface and filamentous algae in the surface waters.

A lake is an **ecosystem**, which consists of the plant and animal communities mentioned above, and the non-living part of the environment (mud, water, minerals, dissolved oxygen, soil and sunlight) on which they depend.

Effect of a limiting factor on population growth

When a limiting factor influences population growth, a sigmoid (S-shaped) curve is created, as shown in Figure 19.13 for a colony of yeast.

You need to be able to identify the **lag**, exponential (**log**), **stationary** and **death** phases on a graph of population growth. A limiting factor such as food takes effect as the population becomes too large for supplies to be sufficient. The population growth rate reduces until births and deaths are equal. At this point, there is no increase in numbers – the graph forms a plateau. As food runs out, more organisms die than are born, so the number in the population drops. This is the death phase.

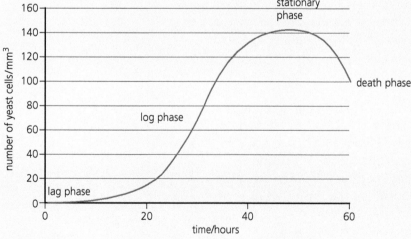

Figure 19.13

Examiner's tip

Make sure you know which is the **lag phase** and which is the **log phase** on a population curve. When you write these terms, make sure your letters 'o' (log) and 'a' (lag) are recognisable.

Cambridge IGCSE Biology Study and Revision Guide Second Edition © Dave Hayward 2016

Abundant food supplies can lead to more people becoming obese; this results in greater demands on health care due to increasing numbers of sufferers of heart disease, diabetes, blindness, etc. In the long term, this may reduce average life expectancy, as poor health becomes a limiting factor.

You need to be able to explain the factors that lead to the different phases shown in Figure 19.13:

- Lag phase – the new population takes time to settle and mature before breeding begins. When this happens, a doubling of small numbers does not have a big impact on the total population size, so the line of the graph rises only slowly with time.
- Log (exponential) phase – there are no limiting factors. Rapid breeding in an increasing population causes a significant increase in numbers. A steady doubling in numbers per unit of time produces a straight line.
- Stationary phase – limiting factors, such as shortages of food, cause the rate of reproduction to slow down and there are more deaths in the population. When the birth rate and death rate are equal, the line of the graph becomes horizontal.
- Death phase – the mortality rate (death rate) is now greater than the reproduction rate, so the population numbers begin to drop. Fewer offspring will live long enough to reproduce. The decline in population numbers can happen because the food supply is insufficient, waste products contaminate the habitat or disease spreads through the population.

Now try this

5 Figure 19.14 shows a population curve for a species of animal colonising a new habitat.
  a (i) Identify phase Y of the curve. [1 mark]
      (ii) Suggest why the population increase in phase X is slow. [2 marks]
  b Identify three factors that limit the size of such a population but do not appear to limit the total human population. [3 marks]

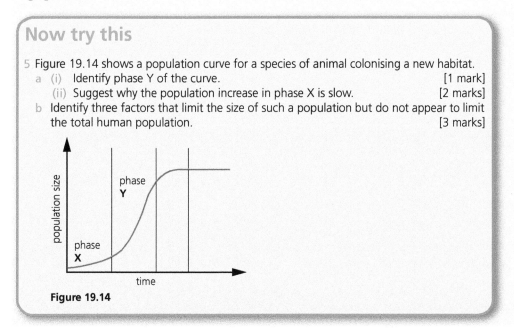

Figure 19.14

Biotechnology and genetic engineering

20

● Key term

Genetic engineering	Changing the genetic material of an organism by removing, changing or inserting individual genes

Biotechnology is the application of biological organisms, systems or processes to manufacturing and service industries. **Genetic engineering** involves the transfer of genes from one organism to (usually) an unrelated species. Both processes often make use of bacteria because of their ability to make complex molecules (e.g. proteins) and their rapid reproduction rate.

> **Now try this**
>
> 1 State why bacteria are very useful organisms in the process of genetic engineering.   [2 marks]

● Use of bacteria in biotechnology and genetic engineering

Bacteria are useful in biotechnology and genetic engineering because:

- they can be grown and manipulated without raising ethical concerns;
- they have a genetic code that is the same as all other organisms, so genes from other animals or plants can be successfully transferred into bacterial DNA.

Bacterial DNA is in the form of a circular strand and also small circular pieces called **plasmids**. Scientists have developed techniques to cut open these plasmids and insert sections of DNA from other organisms into them. When the bacterium divides, the DNA in the modified plasmid is copied, including the 'foreign' DNA. This may contain a gene to make a particular protein.

● Biotechnology

Production of ethanol for biofuels

In Chapter 12, the anaerobic respiration of glucose to alcohol is described as a form of **fermentation**. Micro-organisms that bring about fermentation are using the chemical reaction to produce energy, which they need for their living processes.

Cambridge IGCSE Biology Study and Revision Guide Second Edition   © Dave Hayward 2016

Yeast is encouraged to grow and multiply by providing nutrients such as sugar. Oxygen or air is excluded to maintain an anaerobic process. Ethanol is a waste product. An optimum pH and temperature are maintained for the yeast being cultured. Some countries produce ethanol in this way as a renewable source of energy (**biofuel**) for motor cars, replacing non-renewable petrol.

Bread-making

Yeast is used in bread-making and brewing because of the products produced when it respires. The yeast is mixed with water and sugar to activate it. The mixture is added to flour to make dough. This is left in a warm place to rise. The dough rises because the yeast is releasing carbon dioxide, which gets trapped in the dough. A warm temperature is important because respiration is controlled by enzymes (see Chapter 5). When the dough is cooked, the high temperature kills the yeast and any ethanol formed evaporates. Air spaces are left where the carbon dioxide was trapped. This gives the bread a light texture.

● Common misconceptions

- The production of biofuels and bread using yeast does not involve aerobic respiration. They both involve anaerobic respiration.

The use of pectinase in fruit juice production

When pectinase is added to fruit tissue, it breaks down the tissue, releasing sugars in solution and making the liquid extract transparent (clarifying it). The process occurs faster in warm conditions than in cold conditions because it is controlled by an enzyme.

Use of biological washing powders that contain enzymes

Detergent alone is less effective than biological washing powders, because biological washing powders can contain protease and lipase to remove protein stains and fat/grease from clothes. The enzymes break down the proteins and fats in the clothes to amino acids, fatty acids and glycerol. These are smaller, soluble molecules, which can escape from the clothes and dissolve in the water.

The relatively low temperatures in which enzymes work best make the biological washing powder more efficient because this saves energy (no need to boil water). However, if the temperature is too high the enzymes will be denatured.

The use of lactase to produce lactose-free milk

Lactose is a type of sugar found in milk and dairy products. Some people suffer from **lactose intolerance**, a digestive problem in which the body does not produce enough of the enzyme lactase. As a result, the lactose remains in the gut, where it is fermented by bacteria, causing symptoms such as flatulence (wind), diarrhoea and stomach pains. Many foods contain dairy products, so people with lactose intolerance cannot eat them, or suffer the symptoms described above. However, lactose-free milk is now produced using the enzyme lactase.

The lactase can be produced on a large scale by fermenting yeasts or fungi. The fermentation process is shown in Figure 20.1.

A simple way to make lactose-free milk is to add lactase to milk. The enzyme breaks down lactose sugar into two monosaccharide sugars: glucose and galactose. Both can be absorbed by the intestine.

An alternative, large-scale method is to immobilise lactase on the surface of beads. The milk is then passed over the beads and the lactose sugar is effectively removed. This method avoids having the enzyme molecules in the milk because they remain on the beads.

Penicillium and the production of the antibiotic penicillin

Some micro-organisms, such as the fungus *Penicillium*, produce complex organic compounds called **antibiotics**. The fungus is grown on a large scale (see Figure 20.1), then put under stress by reducing the nutrient supply. This causes it to secrete penicillin, which can be filtered off.

How fermenters are used in the production of penicillin

The fermenter (Figure 20.1) is a large, sterile container with a stirrer, a pipe to add feedstock (molasses or corn-steep liquor) and air pipes to blow air into the mixture. The fungus *Penicillium* is added and the liquid is maintained at around 26 °C and a pH of 5–6. Sterile conditions are essential to prevent

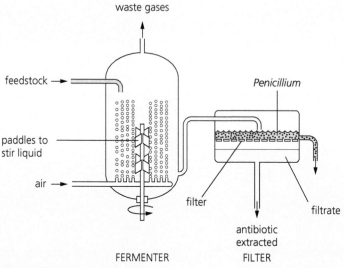

Figure 20.1

'foreign' bacteria or fungi getting into the system, as they can completely disrupt the process. As the nutrient supply diminishes, the fungus begins to secrete antibiotics into the medium. The nutrient fluid containing the antibiotic is filtered off and the antibiotic is extracted by crystallisation or other methods.

● Genetic engineering

You need to learn the definition of *genetic engineering*, given at the start of this chapter, and be able to give examples. Some are listed below:

● The production of human insulin – the human insulin gene is inserted into bacteria. Human insulin does not trigger allergic reactions in the way that animal insulin can, and is acceptable to people with a range of religious beliefs.
● The insertion of genes into crop plants to give them resistance to herbicides (weedkillers) – this enables the farmer to spray the crop to kill weeds, without damaging the crop, and may reduce the use of herbicides.
● The insertion of genes into crop plants to give them resistance to insect pests – the gene enables the plant to produce a poison that makes it resistant to attack by insect larvae.
● The insertion of genes into crop plants to provide additional vitamins – golden rice is a variety of rice that has had a gene for beta-carotene (a precursor of vitamin A) inserted. Golden rice is grown particularly in countries where vitamin A deficiency is a problem and where rice is a staple food. This deficiency often leads to blindness.

Using genetic engineering to put human insulin genes into bacteria

Figure 20.2 shows this process. The steps are numbered on the diagram as in the list below.

1 Human cells with genes for healthy insulin are selected.

2 A chromosome (which is a length of DNA) is removed from the cell.

3 The section of DNA representing the insulin gene is cut from the chromosome using **restriction endonuclease enzyme**. The DNA has sticky ends.

4 A suitable bacterial cell is selected. Some of its DNA is in the form of circular **plasmids**.

5 All the plasmids are removed from the bacterial cell.

6 The plasmids are cut open using the same restriction endonuclease enzyme.

7 The human insulin gene is inserted into the plasmids using **ligase** enzyme, forming **recombinant plasmids**.

8 The plasmids are returned to the bacterial cells (only one is shown in the diagram).

9 The bacterial cells are allowed to reproduce in a fermenter. All the cells produced contain plasmids with the human insulin gene. The bacteria can now produce human insulin on a commercial scale.

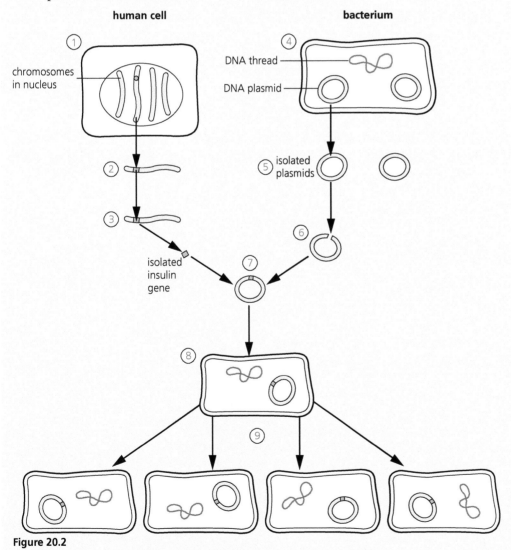

Figure 20.2

● Sample question

Scientists are planning the use of a genetically engineered virus to destroy a population of an amphibian called the cane toad, which is getting out of control in Australia.

1 Define the term *genetic engineering*. [2 marks]

2 State the part of the virus that would carry the modified genetic material. [1 mark]

Student's answer

1 Genetic engineering is changing the genetic material ✔ in an organism by inserting genes ✔ from a different species.
2 The nucleus. ✖

Examiner's comments

The definition given in part 1 is excellent. In part 2, the candidate does not know that a virus lacks a nucleus – it contains DNA or RNA in the form of a strand.

● Advantages and disadvantages of genetically modifying crops

Advantages:

- The aim of most genetic modification is to increase yields through the insertion of genes giving crops herbicide resistance and insect pest resistance. Genetically modified (GM) maize has resistance to pests and herbicides. GM soya has been modified to make it herbicide resistant.
- It is possible to improve the protein, mineral or vitamin content of food. Golden rice has a gene enabling it to produce a precursor of vitamin A. GM soya has an increased nutritional value.
- It is possible to improve the keeping qualities of some products, e.g. the storage properties of GM soya, through modification of its fat molecules using inserted genes. GM tomatoes have had a gene deleted that is responsible for fruit softening, extending their storage life.

Disadvantages:

- The vectors for delivering recombinant DNA contain genes for antibiotic resistance. If these managed to get into potentially harmful bacteria, it might make them resistant to antibiotic drugs.
- GM food could contain pesticide residues or substances that cause allergies.
- The precursor of vitamin A in golden rice could change into other, toxic chemicals once eaten.
- There is a risk of a reduction in biodiversity as a result of the introduction of GM species.
- Subsistence farmers could also be tied to large agricultural suppliers that may then manipulate seed prices.

Examiner's tip

Where a question asks you to state the advantages and disadvantages of a process, make sure you keep your answer balanced. For example, if the question is worth 4 marks, give two advantages and two disadvantages.

Human influences on ecosystems

Key objectives

The objectives for this chapter are to revise:

- definitions of key terms
- how modern technology has resulted in increased food production
- the negative impacts to an ecosystem of large-scale monocultures of crop plants and of intensive livestock production
- the reasons for habitat destruction and the undesirable effects of deforestation
- the sources and effects of pollution of land, water and air
- the need to conserve non-renewable resources (fossil fuels)
- that some resources can be maintained and that some products can be reused or recycled
- how sewage is treated
- why organisms become endangered or extinct and how endangered species can be conserved
- the social, environmental and economic implications of providing sufficient food for an increasing human global population

- the problems which contribute to famine
- the undesirable effects of deforestation on the environment
- the process of eutrophication of water
- the effects of non-biodegradable plastics and acid rain on the environment
- the measures that are taken to reduce sulfur dioxide pollution and reduce the impact of acid rain
- the effects of increases in carbon dioxide and methane concentrations in the atmosphere
- the negative impacts of female contraceptive hormones in water courses
- how forests and fish stocks can be sustained
- what sustainable development requires
- the risks to a species if the population size drops, reducing variation
- the reasons for conservation programmes

Key terms

Sustainable resource	A resource that is produced as rapidly as it is removed from the environment so that it does not run out
Sustainable development	Development providing for the needs of an increasing human population without harming the environment

Food supply

Larger populations require more food, provided by improving methods of agriculture. Modern technology has resulted in increased food production in a number of ways:

- **Agricultural machinery** enables much larger areas of land to be cleared, and makes preparing soil, planting, maintaining and harvesting crops significantly more efficient. The process of farming in general has become more efficient.
- The use of **chemical fertilisers** improves yield. These are mineral salts made on an industrial scale. Examples are ammonium sulfate (for nitrogen and sulfur), ammonium nitrate (for nitrogen) and compound NPK fertiliser for nitrogen, phosphorus and potassium. These are spread on the soil in carefully calculated amounts to provide the minerals that the plants need.
- The use of **insecticides** improves quality and yield. Crops are very susceptible to attack by insect pests. Insecticides combat these attacks, so the crops grow more successfully and show less damage.
- The use of **herbicides** reduces competition with weeds. Weeds are plants that compete with the crop plant for root space, soil minerals and sunlight. Herbicides are chemicals that kill the weeds growing amongst the crop plants.

● **Selective breeding** can be used to improve production by crop plants and livestock. An important part of any breeding programme is the selection of the desired varieties that have particular qualities, such as flavour and disease resistance in plants, and high milk or meat yield or resistance to disease in animals such as cattle, fish and poultry.

● The negative impact to an ecosystem of monocultures

A monoculture is a crop grown on the same land, year after year. Every attempt is made to destroy organisms that feed on, compete with or infect the crop plant. This reduces the number of species in an area and has a negative impact on food chains. The removal of hedges reduces nesting sites for birds and habitats of other organisms. The use of pesticides on monocultures can reduce the number of important insect pollinators, which are required by wild flowers.

● The negative impact to an ecosystem of intensive livestock production

Intensive livestock production is also known as 'factory farming'. Chickens and calves are often reared in large sheds instead of in open fields. Their urine and faeces are washed out of the sheds with water, forming 'slurry'. If this gets into streams and rivers, it supplies an excess of nitrates and phosphates, which can lead to water pollution. Over-grazing can result from too many animals being kept on a pasture. They eat the grass down almost to the roots, and their hooves trample the surface soil into a hard layer. As a result, the rainwater will not penetrate the soil, so it runs off the surface, carrying the soil with it. The soil becomes eroded.

● The problems of world food supplies and the causes of famine

There is not always enough food available in a country to feed the people living there. A severe food shortage can lead to famine. Food may have to be brought in (imported). Fresh food can have a limited storage life, so it needs to be transported quickly or treated to prevent it going rotten. Methods to increase the life of food include transport in chilled containers and picking the produce before it is ripe. When it has reached its destination, it is exposed to chemicals such as plant auxins to bring on the ripening process. The use of aeroplanes to transport food is very expensive. The re-distribution of surplus food from first-world countries to a poorer one can have a detrimental effect on that country's local economy by reducing the value of food grown by local farmers. Some food grown by countries with large debts may be exported as cash crops, even though local people desperately need the food. Other problems that can result in famine include:

● climate change and natural disasters such as flooding or drought;
● pollution;
● a shortage of water through its use for other purposes, the diversion of rivers and building dams to provide hydroelectricity;

Cambridge IGCSE Biology Study and Revision Guide Second Edition   © Dave Hayward 2016

- eating next year's seeds through desperation for food;
- poor soil and lack of inorganic ions or fertiliser;
- desertification caused by soil erosion, as a result of deforestation;
- poverty – the lack of money to buy seeds, fertiliser, pesticides or machinery;
- war, which can make it too dangerous to farm or which removes labour;
- urbanisation (building on farm land);
- an increasing population;
- pest damage or disease;
- poor education of farmers and outmoded farm practices;
- the destruction of forests, so there is nothing to hunt and no food to collect.

Habitat destruction

There are three key reasons for habitat destruction:

- an increased area of land is needed for food crop growth, livestock production and housing as the human population increases;
- the extraction of more natural resources, as we need more raw materials for the manufacturing industry and greater energy supplies;
- marine pollution – marine habitats are becoming contaminated with human debris, including untreated sewage, agricultural fertilisers, pesticides, non-biodegradable plastics and waste oil.

The effects of altering food webs and food chains on habitats

If human activity causes one population of organisms to die or disappear, a food web or food chain becomes unbalanced. For example, an increase in herbivores due to the over-hunting of a carnivore may result in the over-grazing of land. Once the plants have been removed, the soil is vulnerable to erosion because there are no roots to absorb water or to hold the soil together. The habitat would then be destroyed.

Undesirable effects of deforestation

Deforestation is the removal of large areas of forest to provide land for farming and roads, and to provide timber for building, furniture and fuel. The removal of large numbers of trees results in habitat destruction on a massive scale, which can have the follows results:

- Animals living in the forest lose their homes and sources of food; species of plant become extinct as the land is used for other purposes such as agriculture, mining, housing and roads.
- Soil erosion is more likely to happen, as there are no roots to hold the soil in place. The soil can end up in rivers and lakes, destroying habitats there.
- Flooding becomes more frequent, as there is no soil to absorb and hold rainwater. Plant roots rot and animals drown, destroying food chains and webs.
- Carbon dioxide builds up in the atmosphere, as there are fewer trees to photosynthesise, increasing global warming. Climate change affects habitats.

Now try this

1 Figure 21.1 shows the area of tropical rainforest deforested annually in five different countries, labelled A to E.

  a (i) Which of the countries shown has the largest area deforested annually?
         [1 mark]

      (ii) Which of the countries shown has 600 000 hectares of rainforest removed each year? [1 mark]

     (iii) In another country (F), 550 000 hectares are deforested annually. Plot this on a copy of the figure.
         [1 mark]

  b (i) Country E has a total of 9 000 000 hectares of tropical rainforest remaining. How long will it be before it is all destroyed, if the present rate of deforestation continues?
         [1 mark]

      (ii) State two reasons why tropical rainforests are being destroyed by humans. [2 marks]

     (iii) After deforestation has taken place, soil erosion often occurs rapidly. Suggest two ways in which this may occur. [2 marks]

  c Tropical rainforests reduce the amount of carbon dioxide and increase the amount of oxygen in the atmosphere. Explain why both these voccurrences are important to living organisms. [2 marks]

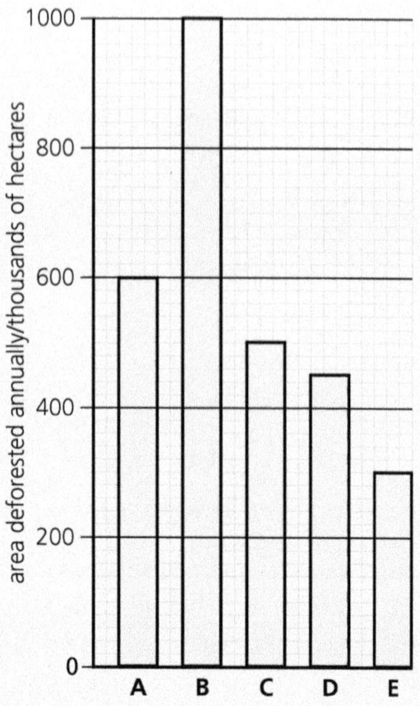

Figure 21.1

The undesirable effects of deforestation on the environment

- The reduction of habitats or food sources for animals, which can result in their extinction. Animal and plant diversity is reduced and food chains are disrupted.
- The loss of plant species and their genes that may be important for medical use or genetic engineering in the future.
- The loss of roots to hold soil together, which can result in soil erosion and leaching of minerals. Desertification can eventually occur.
- The loss of roots and soil can lead to flooding and mudslides. Lakes can become silted up.
- The leaching of nutrients into lakes and rivers, which can lead to eutrophication.
- Less carbon dioxide is absorbed from the atmosphere, increasing the greenhouse effect.
- Less oxygen is produced, so atmospheric oxygen levels can drop.
- Less transpiration, which can lead to reduced rainfall.

The sources and effects of pollution

Land pollution

Some **insecticides** are non-biodegradable and stay in the environment for a long time. For example, DDT has been used to kill mosquitoes to reduce the spread of malaria. However, because it does not break down, it enters water systems such as lakes, where it is absorbed into plankton. Bioaccumulation occurs: the top carnivores suffer from its toxicity. Some insecticides are

Cambridge IGCSE Biology Study and Revision Guide Second Edition © Dave Hayward 2016

non-specific: when applied to kill an insect pest, they also kill all the other insects that are exposed to it. This may include useful insects (e.g. bees, which are needed to pollinate crops). Food webs can be affected, threatening the extinction of top carnivores such as birds of prey.

Herbicides are used to kill weeds in a crop field so that competition can be reduced and therefore crop yield can be increased. However, herbicides may also kill rare plant species near the field being sprayed.

Nuclear fallout can be the result of a leak from a nuclear power station, or from a nuclear explosion. Radioactive particles are carried by the wind or water and gradually settle in the environment. If the radiation has a long half-life, it remains in the environment and is absorbed by living organisms. The radioactive material bioaccumulates in food chains and can cause cancer in top carnivores.

Water pollution

Chemical waste such as **heavy metals** (mercury, nickel, etc.) and oil can cause serious pollution. Some chemicals may be dumped (or enter water systems through leaching) in low concentrations, at which levels they are not toxic. However, bioaccumulation occurs if they enter and pass along a food chain. Animals, including humans, at the top of the food chain accumulate high concentrations of the chemical, which is now toxic. Poisons such as **mercury** damage the central nervous system and can lead to death.

When **oil** is dumped into water it can form a surface layer, coating animals such as birds that feed in the water. When the birds try to clean their feathers they swallow the oil, which poisons them. Oil also disrupts food chains.

Discarded rubbish can result in disease and pollution. It attracts vermin, which are vectors of disease. A lot of rubbish ends up in the sea, causing severe problems for marine animals.

Untreated sewage contains disease organisms, which may get into drinking water and spread diseases such as typhoid and cholera. It also attracts vermin, which are vectors of disease.

Fertilisers – it is very tempting for farmers to increase the amount of fertilisers applied to crops to try and increase crop yields. However, this can lead to the eutrophication of rivers and lakes. Overuse of fertilisers can also lead to the death of the plants. High concentrations of the fertiliser around plant roots can cause the roots to lose water by osmosis. The plant then wilts and dies.

Air pollution by carbon dioxide and methane

Levels of **carbon dioxide** in the atmosphere are influenced by natural processes and by human activities. The main source of pollution that changes the equilibrium (balance) is the combustion of fossil fuels (coal, oil and gas). An increase in the levels of carbon dioxide in the atmosphere is thought to contribute to global warming. Carbon dioxide forms a layer in the atmosphere, which traps heat radiation from the Sun.

Methane also acts as a greenhouse gas. It is produced by the decay of organic matter in anaerobic conditions, such as in wet rice fields and in the stomachs of animals, e.g. cattle and termites. It is also released from the ground during the extraction of oil and coal.

The build-up of greenhouse gases causes a gradual increase in the atmospheric temperature, known as the **enhanced greenhouse effect**. This can:

- melt polar ice caps, causing flooding of low-lying land;
- change weather conditions in some countries, by increasing flooding or reducing rainfall and thus changing arable (farm) land to desert; extreme weather conditions become more common;
- cause the extinction of some species that cannot survive in raised temperatures.

Cambridge IGCSE Biology Study and Revision Guide Second Edition © Dave Hayward 2016

● Sample question

Draw one line from each pollutant in the left column to an environmental effect it might have.

[6 marks]

pollutant		environment effect
untreated sewage		reduces the oxygen-carrying capacity of blood
carbon monoxide		can result in acid rain
sulfur dioxide		can poison top carnivores
insecticides		can cause global warming
methane		can cause mutations
ionising radiation		can lead to water-borne diseases such as cholera

Student's answer

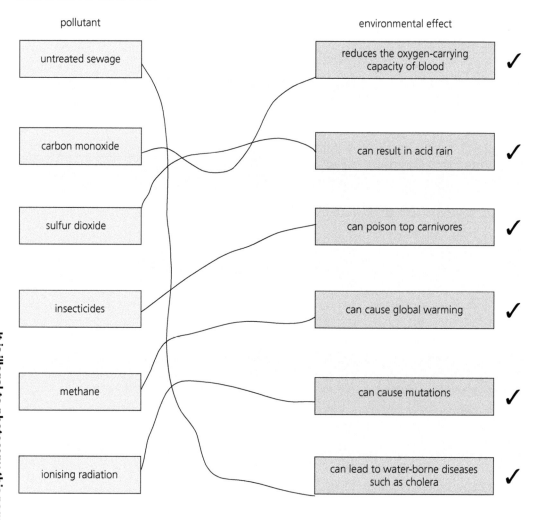

Cambridge IGCSE Biology Study and Revision Guide Second Edition © Dave Hayward 2016

Examiner's comments

● Eutrophication

Sewage and fertilisers both contain high levels of nutrients such as nitrates and other ions. The nitrates act as fertilisers for producers, e.g. algae, which grow and die more rapidly. Decomposers such as bacteria feed on the dead organic matter and reproduce rapidly, using up dissolved oxygen in respiration. Animals in the water system die because of a lack of dissolved oxygen for aerobic respiration. Figure 21.2 shows a flowchart of eutrophication.

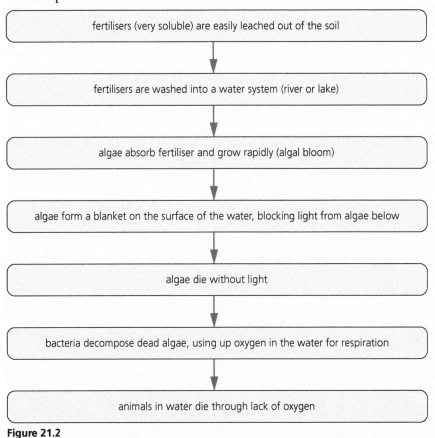

Figure 21.2

Cambridge IGCSE Biology Study and Revision Guide Second Edition © Dave Hayward 2016

Now try this

2 Figure 21.3a shows part of a river into which sewage is pumped. The river water flows from W to Z, with the sewage being added at X. Some of the effects of adding sewage to the river are shown in Figure 21.3b.

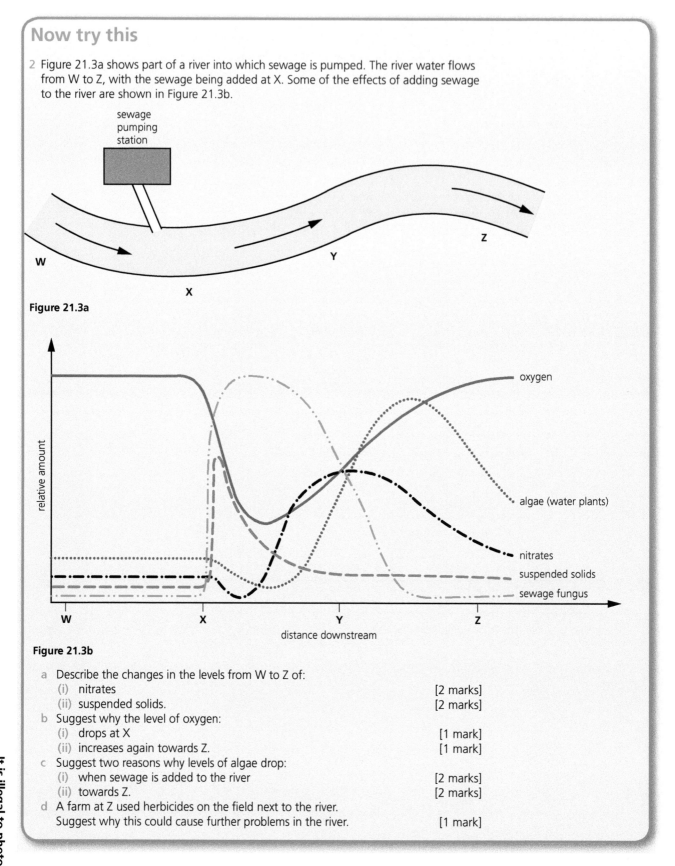

Figure 21.3a

Figure 21.3b

a Describe the changes in the levels from W to Z of:
   (i) nitrates   [2 marks]
   (ii) suspended solids.   [2 marks]
b Suggest why the level of oxygen:
   (i) drops at X   [1 mark]
   (ii) increases again towards Z.   [1 mark]
c Suggest two reasons why levels of algae drop:
   (i) when sewage is added to the river   [2 marks]
   (ii) towards Z.   [2 marks]
d A farm at Z used herbicides on the field next to the river.
   Suggest why this could cause further problems in the river. [1 mark]

Cambridge IGCSE Biology Study and Revision Guide Second Edition © Dave Hayward 2016

Non-biodegradable plastics

Plastics that are non-biodegradable are not broken down by decomposers when dumped in landfill sites or left as litter. This means that they remain in the environment, taking up valuable space or causing visual pollution. Discarded plastic bottles can trap small animals, and nylon fishing lines and nets can trap birds and mammals such as seals and dolphins. As the plastic gradually breaks up into smaller fragments in the sea, it can clog up the gills of fish or get trapped in their stomachs, making them ill.

Acid rain

The main causes of acid rain are processes that release sulfur dioxide and oxides of nitrogen into the atmosphere. These include the burning of fossil fuels, such as coal and gas, by power stations and the combustion of petrol in car engines.

The oxides of sulfur and nitrogen dissolve in the water vapour in clouds, forming acids. When it rains, the rain is acidic.

Problems caused by acid rain include the following:

- damage to plant leaves, eventually killing the plants – whole forests of pine trees have been destroyed by acid rain;
- the acidification of lakes – as the water becomes more acidic, some animals such as fish cannot survive and fish stocks are destroyed;
- an increased risk of asthma attacks and bronchitis in humans;
- corrosion of stonework on buildings;
- the release into soil of soluble aluminium ions that are toxic to fish when washed into lakes.

Ways of reducing the incidence of acid rain include:

- changing the types of power stations that generate electricity from coal and oil to gas or nuclear power, or using more renewable energy sources such as wind;
- using 'scrubbers' in power station chimneys – these remove most of the sulfur dioxide present in the waste gases.

Common misconceptions

- Car engines do not make large amounts of sulfur dioxide, but they are responsible for producing large amounts of oxides of nitrogen, carbon dioxide and carbon monoxide.

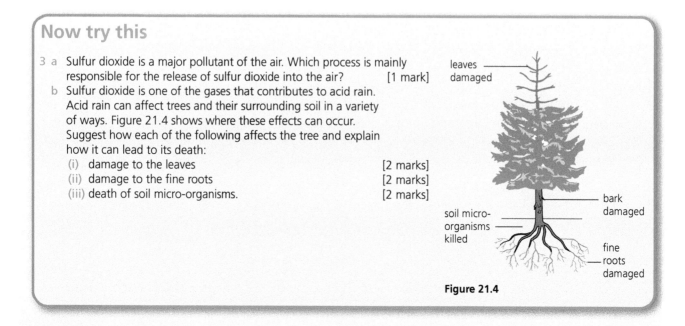

Now try this

3 a Sulfur dioxide is a major pollutant of the air. Which process is mainly responsible for the release of sulfur dioxide into the air? [1 mark]
  b Sulfur dioxide is one of the gases that contributes to acid rain. Acid rain can affect trees and their surrounding soil in a variety of ways. Figure 21.4 shows where these effects can occur. Suggest how each of the following affects the tree and explain how it can lead to its death:
   (i) damage to the leaves [2 marks]
   (ii) damage to the fine roots [2 marks]
   (iii) death of soil micro-organisms. [2 marks]

leaves damaged

bark damaged

soil micro-organisms killed

fine roots damaged

Figure 21.4

● The greenhouse effect and climate change

The atmosphere acts like the glass in a greenhouse. It lets in light and heat from the Sun but reduces the amount of heat that escapes. Carbon dioxide and methane are gases that absorb a lot of infrared radiation. So, if the concentration of either of these gases were to increase, the 'greenhouse effect' would be enhanced and the Earth would get warmer.

Since the Industrial Revolution, we have been burning 'fossil fuels' derived from coal and petroleum and releasing extra carbon dioxide into the atmosphere. As a result, the concentration of carbon dioxide has increased and is likely to go on increasing as we burn more and more fossil fuel. Methane is produced by anaerobic respiration, e.g. in cows' stomachs and in paddy fields where rice is cultivated. Increased cattle and rice production therefore increase methane levels in the atmosphere.

Examples of climate change caused by the enhanced greenhouse effect have already been listed earlier in this chapter.

● Pollution by contraceptive hormones

When women use the contraceptive pill, the hormones in it (oestrogen or progesterone, see Chapter 16) are excreted in urine, become present in sewage and end up in water systems such as rivers and lakes. The hormones affect aquatic organisms as they enter food chains. For example, male frogs and fish can become 'feminised' (they can start producing eggs in their testes instead of sperm). This causes an imbalance between the sexes, with more females than males. Drinking water that is extracted from rivers where water from treated sewage has been recycled can contain the hormones. This has been shown to reduce the sperm count in men, causing a reduction in fertility.

Conservation

The definition of a *sustainable resource* is given at the start of this chapter.

Conservation of resources

Some natural resources (the materials we take from the Earth) are non-renewable. For example, fossil fuels such as coal take millions of years to form. Increasing demands for energy are depleting these resources. One way of conserving these resources is to increase the use of renewable energy (wind farms, solar power, hydroelectric power, production of biofuels, etc.). Another is to improve the efficiency of energy use (better insulation, smaller car engines, more public transport, etc.). Trees can be grown specifically for fuel, then replanted as they are cut down. In this way, the greenhouse effect is not increased and habitats can be maintained when tree felling is carefully managed.

Some resources, such as forests and fish stocks, can be maintained with careful management. This may involve replanting land with new seedlings as mature trees are felled and controlling the activities of fishermen operating where fish stocks are being depleted.

Recycling and reuse

As minerals and other resources become scarcer, they also become more expensive. It then pays to use them more than once. The recycling of materials can reduce the amount of energy used in manufacturing. In turn, this helps to conserve fuels and reduce pollution. For example, producing aluminium alloys from scrap uses only 5% of the energy that would be needed to make them from aluminium ores. Other metals such as iron, copper and lead are also recycled.

Manufacturing glass bottles uses about three times more energy than if they were collected, sorted, cleaned and reused. Recycling the glass from bottles does not save energy but does reduce the demand for sand used in glass manufacture.

Polythene waste is now also recycled. The plastic is used to make items such as car seat covers and sports shoes.

Waste paper can be pulped and used again, mainly for making paper and cardboard.

Recycling sewage

Sewage is mainly water, contaminated with organic material, solids, bacteria and minerals such as phosphates. In places where water is in short supply, the sewage is treated to provide water that is clean enough to drink. Any treated effluent that is returned to a water system such as a river will not cause problems such as eutrophication.

Figure 21.5 describes the treatment of sewage to provide clean water.

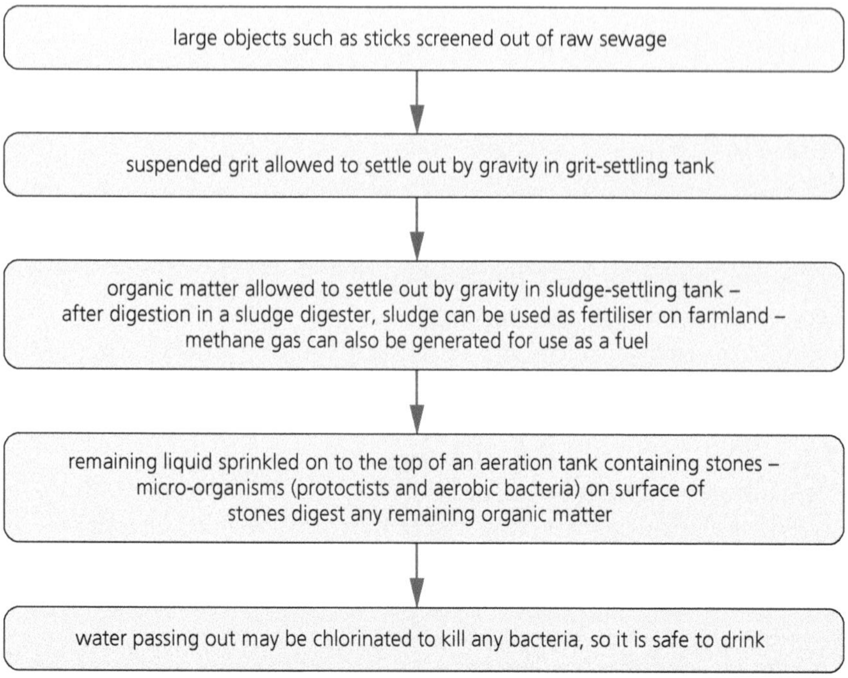

```
┌─────────────────────────────────────────────────────────────┐
│     large objects such as sticks screened out of raw sewage   │
└─────────────────────────────────────────────────────────────┘
                              │
                              ▼
┌─────────────────────────────────────────────────────────────┐
│  suspended grit allowed to settle out by gravity in grit-settling tank │
└─────────────────────────────────────────────────────────────┘
                              │
                              ▼
┌─────────────────────────────────────────────────────────────┐
│     organic matter allowed to settle out by gravity in sludge-settling tank – │
│  after digestion in a sludge digester, sludge can be used as fertiliser on farmland – │
│        methane gas can also be generated for use as a fuel     │
└─────────────────────────────────────────────────────────────┘
                              │
                              ▼
┌─────────────────────────────────────────────────────────────┐
│  remaining liquid sprinkled on to the top of an aeration tank containing stones – │
│       micro-organisms (protoctists and aerobic bacteria) on surface of │
│              stones digest any remaining organic matter        │
└─────────────────────────────────────────────────────────────┘
                              │
                              ▼
┌─────────────────────────────────────────────────────────────┐
│ water passing out may be chlorinated to kill any bacteria, so it is safe to drink │
└─────────────────────────────────────────────────────────────┘
```

Figure 21.5

● Endangering species and causing their extinction

Many species of animals and plants are becoming endangered or are in danger of extinction. This is because of factors such as climate change, habitat destruction, hunting, pollution and the introduction of other species.

● Conservation of species

Conservation can sometimes be achieved by the monitoring and protection of the species. Many organisations monitor species numbers, so that conservation measures can be taken if they decline significantly.

The conservation of habitats is equally as important as the conservation of individual species. If habitats are lost, so are the species that live in them, so habitat destruction poses the greatest threat to the survival of species. A habitat may be conserved by:

● using laws to protect the habitat;
● using wardens to protect the habitat;
● reducing or controlling public access to the habitat;
● controlling factors, such as water drainage and grazing, that may otherwise contribute to destruction of the habitat.

Education plays an important role in helping local communities to understand why species need to be conserved.

Provided a species has not become totally extinct, it may be possible to boost its numbers by breeding in captivity and releasing the animals back into the environment.

Seed banks are a way of protecting plant species from extinction. They include seed from food crops and rare species. They act as gene banks.

Cambridge IGCSE Biology Study and Revision Guide Second Edition © Dave Hayward 2016

● Sustaining forests and fish stocks

The definition of *sustainable development* is given at the start of this chapter.

It is possible to sustain resources such as forests and fish stocks using education, legal quotas and restocking.

Education

Education usually involves sharing information with local communities about the need for conservation. Once they understand its importance, the environment they live in is more likely to be cared for and the species in it protected.

In tropical rainforests, it has been found that the process of cutting down the trees damages twice as many trees next to them, and dragging the trees out of the forest creates more damage. Educating people about alternative ways of tree felling, reduction of wastage and the selection of species of trees to be felled makes the process more sustainable and helps to conserve rarer species.

Legal quotas

In Europe, the Common Fisheries Policy is used to set quotas for fishing, to manage fish stocks and to help protect species that are becoming endangered through over-fishing. Quotas are set for each species of fish taken commercially and also for the size of fish. This is to allow fish to reach breeding age and to maintain or increase their populations.

The Rainforest Alliance has introduced a scheme called *SmartLogging*. This is a certification service that demonstrates that a logging company is working legally and in a sustainable way to protect the environment. The timber can be tracked from where it is felled to its final export destination and its use in timber products.

Restocking

Where populations of a fish species are in decline, their numbers may be conserved by a restocking programme. This involves breeding fish in captivity then releasing them into the wild.

● Sustainable development

This is a complex process requiring the management of conflicting demands. As the world's population grows, so does the demand for the extraction of resources from the environment. However, this needs to be carried out in a controlled way to prevent environmental damage and strategies need to be put in place to ensure habitats and species diversity are not threatened. Planning and co-operation need to be applied at local, national and international levels. This is to make sure that everyone involved with the process is aware of the potential consequences of the process on the environment, and that appropriate strategies are put in place, and adhered to, to minimise any risk.

If the population of a species drops, the range of variation within the species drops, making it less able to adapt to environmental change. The species could, therefore, be threatened with extinction. When animal populations fall, there is less chance of individuals finding each other to mate.

Cambridge IGCSE Biology Study and Revision Guide Second Edition © Dave Hayward 2016

● Conservation programmes

Conservation programmes are set up for a number of reasons, which are outlined below.

Reducing extinction

Conservation programmes strive to prevent extinction. Once a species becomes extinct, its genes are lost forever, so we are also likely to deprive the world of genetic resources. We could deprive ourselves of the beauty and diversity of species, as well as potential sources of valuable products such as drugs.

Protecting vulnerable environments

Conservation programmes are often set up to protect threatened habitats so that rare species living there are not endangered. There are a number of organisations involved with habitat conservation in Britain, e.g. Natural England.

Maintaining ecosystem functions

There is a danger of destabilising food chains if a single species in that food chain is removed. For example, in lakes containing pike as the top predators, over-fishing can result in smaller species of carnivorous fish, such as minnows, increasing in numbers. They eat zooplankton. If the minnows eat the majority of the zooplankton population, it leaves no herbivores to control algal growth, which can lead to eutrophication. To prevent such an event happening, the ecosystem needs to be maintained by controlling the numbers of top predators removed or by regular restocking.

Ecosystems can also become unbalanced if the nutrients they rely on are affected in some way. This may be due to the unregulated removal of materials that affect food chains indirectly because of changes to nutrient cycles.

The term *ecosystem services* can be defined as the benefits people obtain from ecosystems, whether they are natural or managed. Humans are affecting ecosystems on a large scale because of the growth in the population and changing patterns of consumption. About 40% of the Earth's land surface area is taken over by some form of farmed land. Crops are grown for food, extraction of drugs (both legal and illegal) and the manufacture of fuel. Crop growth has major impacts on ecosystems, causing the extinction of many species and reducing the gene pool.

Cambridge IGCSE Biology Study and Revision Guide Second Edition © Dave Hayward 2016

Answers

Chapter 1

1 Nutrition, respiration, movement, excretion.
2 A, fish; B, amphibian; C, mammal; D, bird; E, reptile.
3 One possible mnemonic is Pink Panthers Find People Annoying.
4 a Suitable features, as listed in Figure 1.2.
   b Exoskeleton, segmented body, jointed limbs.
5

Leaf	1a	1b	2a	2b	3a	3b	4a	4b	5a	5b	Name of tree
B		✔				✔				✔	*Quercus*
C		✔			✔			✔			*Ilex*
D		✔		✔			✔				*Fraxinus*
E		✔		✔	✔						*Aesculus*
F	✔		✔								*Magnolia*

Chapter 2

1 Drawings should be suitably labelled.
2 A, upper epidermis. Two details from:
   - a single layer of cells;
   - produces or secretes wax or cuticle;
   - to make leaf waterproof;
   - reference to transparent nature of cells;
   - to allow light to pass through;
   - reference to acting as a barrier against bacteria or fungi.

   B, palisade mesophyll. Two details from:
   - cells are very long or columnar;
   - cells contain chloroplasts or chlorophyll;
   - reference to photosynthesis.

   C, spongy mesophyll. Two details from:
   - cells are rounded;
   - air spaces are present between cells;
   - reference to photosynthesis;
   - reference to gaseous exchange.

   D, guard cells. Two details from:
   - present in pairs;
   - surround a stomatal pore;
   - control the opening or closing of the pore;
   - reference to gaseous exchange;
   - reference to control of transpiration.

3 Width from A to B = 7.5 cm (or 75 mm), and magnification = observed size / actual size
   So, actual size = observed size / magnification = 7.5 / 2.5 = 3.0 cm (or 30 mm).

Chapter 3

1 a (i) The volume of water in the dish decreased.
     (ii) The volume of salt solution in the potato increased.
   b (i) Osmosis.
     (ii) Three points from:
        - there was a higher concentration of water in the dish than in the potato (or there was a higher concentration of salt in the potato than in the dish);
        - so water moved into the potato;
        - from a high concentration of water to a lower concentration of water;
        - by osmosis.
     (iii) Root hairs, or in the roots.
     (iv) Osmosis enables the plant to absorb water to maintain cell turgidity (or to replace water lost by transpiration).
2 a Two points from:
     - reference to large numbers;
     - cells have a large surface (area);
     - mitochondria are present to provide energy.
   b (i) Two points from:
        - absorption of a substance into a cell or across a membrane;
        - against or up a concentration gradient;
        - reference to the process using energy.
     (ii) Active transport requires energy.

Chapter 4

1

Biological molecule	Chemical elements present	Sub-unit(s)	Examples
Carbohydrate	Carbon, hydrogen, oxygen	Glucose	Starch, glycogen, cellulose, sucrose
Fat/oil (oils are liquid but fats are solid at room temperature)	Carbon, hydrogen, oxygen (but lower oxygen content than carbohydrates)	Fatty acids and glycerol	Vegetables oils, e.g. olive oil; animal fats, e.g. cod liver oil, waxes
Protein	Carbon, hydrogen, oxygen, nitrogen, sometimes sulfur or phosphorus	Amino acids (about 20 different forms)	Enzymes, muscle, haemoglobin, cell membranes

Chapter 5

1

Name of enzyme	Substrate (what the enzyme works on)	End product(s)	Other details (e.g. where reaction happens, optimum pH)
Salivary amylase	Starch	Maltose, glucose	Mouth, pH 6.8
Protease	Protein	Amino acids	Stomach, pH 2; duodenum, pH 9
Pancreatic lipase	Fat	Fatty acids, glycerol	Duodenum, pH 9

2 a The *x*-axis should be temperature, with suitable scale (0–50 or 15–50 °C) and labelled with units. The *y*-axis should be time from 0 to 35 or from 0 to 40 minutes and labelled with units.
All points plotted accurately and line drawn to link points on the graph.

b (i) One way is to take small samples of the mixture at regular time intervals and test these with iodine solution. When all the starch has been digested, the mixture will not turn blue/black (an orange/brown colour would appear).

(ii) 35 °C.

(iii) As the temperature is increased from 15 °C to 35 °C, the rate of starch digestion increases. From 35 °C to 50 °C, the rate of starch digestion decreases.

c (i) The starch would be digested very quickly (in about three minutes) because exposure to the low temperature would not affect the enzyme, just slow down its action.

(ii) The starch would not be digested at all, because exposure to the high temperature would denature the enzyme. This effect is permanent.

3 a (i) Blood contains proteins, so trypsin will be needed to digest them.

(ii) Amino acids.

(iii) One suitable use such as: for growth; the production of enzymes; the formation of muscle or cell membranes.

b The mosquitoes would starve to death, because they would not be able to digest the proteins in the blood they feed on.

c Boiling it or heating it; exposing it to extreme pH.

Chapter 6

1

Part of leaf	Results of starch test	Reason
A	Negative – orange/brown	No light to photosynthesise, so no starch produced
B	Positive – blue/black	Light and chlorophyll to photosynthesise, so starch is produced
C	Negative – orange/brown	No chlorophyll to photosynthesise, so no starch produced
D	Negative – orange/brown	No light or chlorophyll to photosynthesise, so no starch produced

2 a (i) Curve A, because the rate of sugar production is higher owing to a higher rate of photosynthesis.

(ii) Chloroplasts contain chlorophyll, which traps sunlight.

(iii) One answer from:
- thickness of the leaf;
- transparency of the leaf;
- angle of the leaf to the light.

b (i) Starch is insoluble, so it stays where it is stored.

(ii) Iodine solution.

3 a 300 / 150 × 100 = 200%.

b Container A – one point, such as:
- depletion of salts or nutrients;
- release of seeds;
- effects of disease;
- shortage of carbon dioxide;
- end of the plant's life cycle.

Container B – one point, such as:
- nutrients still available;
- photosynthesis or growth;
- food stored;
- plenty of carbon dioxide available.

c Container C. One point from:
- least photosynthesis happening;
- respiration is happening faster than photosynthesis;
- the plant has died, so bacteria feed on it, using up oxygen.

d Green or blue. One point from:
- little or no photosynthesis is happening;
- no light absorbed by chlorophyll;
- no sugars (or starch) made.

4 One mark for each statement (as in the table on p. 30).

Chapter 7

1 Any useful mnemonic, such as Can Flowers Pollinate Very Much (in) Freezing Winters?

2 a Z.   **b** Y.   **c** X.

3 a Incisor, canine, premolar and molar tooth labelled in the correct places.

b (i) Biting or cutting food.

(ii) Chewing, grinding or crushing food.

c (i) Enamel.

(ii) Mineral: calcium.
Vitamin: D.

Cambridge IGCSE Biology Study and Revision Guide Second Edition   © Dave Hayward 2016

(iii) Three points from:
- bacteria feed on sugar from food left on the teeth;
- bacteria produce acid;
- acid attacks or dissolves the enamel;
- dentine is softer, so it breaks down more quickly;
- this results in a hole in the enamel, exposing the pulp cavity.

Chapter 8

1 a Labels should include cell wall, membrane, nucleus, cytoplasm and sap vacuole.
  b Chloroplast.
  c (i) Osmosis.
    (ii) Diffusion or active transport (or active uptake).
2 a A, stoma or stomatal pore; B, guard cell; C, epidermal cell or epidermis.
  b *C. fistula* has stomata on upper surface and *B. monandra* has stomata on lower surface. *B. monandra* has more stomata than *C. fistula* (or you could compare the figures).
  c (i) Three points from:
    - colour change is due to water loss;
    - water lost through the stomata;
    - stomata are open during the day;
    - reference to transpiration.
    (ii) One point from:
    - stomata or guard cells are closed at night;
    - there is no transpiration or no water loss at night.
  d Three points from:
    - reference to xylem or tracheids;
    - water enters xylem vessel through pits in walls;
    - reference to osmosis (for absorption of water);
    - reference to transpiration stream or transpiration pull;
    - reference to capillary action or to the cohesion of water molecules;
    - reference to root pressure.
  e (i) The rate will decrease. One point from:
    - there is a smaller gradient;
    - reference to diffusion.
    (ii) The rate will increase. One point from:
    - more energy or heat for evaporation;
    - water evaporates more quickly at higher temperatures;
    - warm air can hold more water vapour than cold air.

Chapter 9

1 a (i) Labels correctly placed for the pulmonary artery, pulmonary vein, vena cava and aorta.
    (ii) Labels correctly placed for the left atrium, left ventricle, right atrium and right ventricle.

(iii) Tricuspid valve (between right atrium and right ventricle); bicuspid valve (between left atrium and left ventricle).
  b Blood leaving the right ventricle has more carbon dioxide and less oxygen than blood entering the left atrium.
2 Correctly labelled diagram and shading.

Chapter 10

1

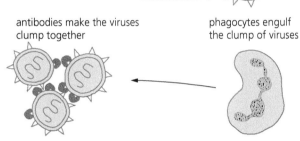

Chapter 11

1 Wall of alveolus – one cell thick (or very thin), so that diffusion happens quickly.
Moist surface – allows oxygen to dissolve, making diffusion faster.
Blood flow – blood is moving, so a concentration gradient is maintained for oxygen and carbon dioxide.
Wall of capillary – one cell thick (or very thin) so that diffusion happens quickly.
Red blood cells – contain haemoglobin to transport oxygen away from the lungs.
2 a (i) Inspired air contains more oxygen, less carbon dioxide and less water vapour than expired air.
    (ii) Three features from:
    - the wall of each alveolus is one cell thick (or very thin);
    - there is a moist surface to the alveoli;
    - there are large numbers of alveoli;
    - the air in each alveolus is constantly being replaced.
  b (i) The release of energy by cells without the use of oxygen.
    (ii) In muscle cells.
3 Breathing in:
- external intercostal muscles contract;
- ribcage moves upwards and outwards;
- the diaphragm muscles contract;
- the diaphragm moves down;
- the volume of the thorax increases;

Cambridge IGCSE Biology Study and Revision Guide Second Edition © Dave Hayward 2016

- the air pressure in the thoracic cavity reduces;
- air rushes into the lungs through the mouth or nose.

For breathing out, the descriptions are the opposite of breathing in.

Chapter 12

1

Type of respiration	Requirement(s)	Products(s)	Amount of energy released
Aerobic	Oxygen, glucose	Water, carbon dioxide, energy	38 molecules of ATP
Anaerobic	Glucose	Lactic acid, energy	1 molecule of ATP

Chapter 13

1 a X, ureter; Y, bladder; Z, urethra.
  b Renal artery.
  c Blood going to the kidneys contains more urea.
    Blood going to the kidneys contains more oxygen.

Chapter 14

1 a (i) Motor neurone.
    (ii) Two features from:
       - presence of motor end plates;
       - the cell body is at the beginning of the cell;
       - the cell body has dendrites on;
       - there is no dendron (only an axon).
    (iii) Peripheral nervous system.
  b Cytoplasm – two points, such as:
    - is elongated;
    - passes electrical signals along;
    - connects different parts of the body;
    - is modified to form dendrites.
    Myelin sheath – two points, such as:
    - acts as insulating material;
    - so prevents leakage of electrical signal from axon;
    - allows faster transmission of impulses.
  c (i) Stimulus, receptor, co-ordinator, effector, response.

    (ii)
Part of sequence	Part in pupil reflex
Effector	Iris (muscle)
Receptor	Retina or rods or cones
Response	Pupil changes diameter or iris muscles contract
Stimulus	Bright light or change in light intensity

2 Suitably labelled diagrams.
3 • Ciliary muscles contract;
  • the suspensory ligaments become relaxed;
  • so tension is removed from the lens;

- the lens becomes more convex;
- so light is focused more strongly.

4
Feature	Nervous control	Hormonal control
Speed	Extremely rapid	Slow(er)
Pathway	Neurones	Bloodstream
Nature of 'impulse'	Electrical	Chemical
Origin	Sense organ/brain	Endocrine gland

5 Glucose, pancreas, secretion, glycogen, insulin, liver.
6 a (i) Positive phototropism.
    (ii) Three points from:
       - the coleoptiles have been exposed to one-sided light;
       - auxins have been produced by the tip;
       - and have passed into the block;
       - auxins have passed from the block to the cut coleoptile;
       - more auxin accumulates on the shaded side of the coleoptile;
       - causing more growth on the shaded side.
  b A, taller and growing vertically upwards.
    B, taller and bending towards the light.
    C, taller and growing vertically upwards.

Chapter 15

1 a (i) Depressant: a drug that acts on the central nervous system, leading to relaxation and sleep or unconsciousness.
    Addictive: causes the development of dependence on a drug.
    (ii) Two long-term effects, such as:
       - liver damage (cirrhosis);
       - brain damage;
       - alcoholism;
       - stomach ulcers;
       - obesity.
  b His or her performance will reduce as the amount of alcohol drunk increases. Reactions will be slower owing to decreased co-ordination, resulting in a greater risk of accidents.

Chapter 16

1 Three points (in sequence) from:
   - the wall of the uterus contracts;
   - the amniotic sac bursts;
   - amniotic fluid passes out through the vagina;
   - the cervix dilates;
   - the baby passes out through the cervix and vagina.
2 a A, umbilical cord; B, vagina.
  b Three functions from:
     - transfers oxygen from mother to fetus;
     - transfers nutrients (or named nutrients) from mother to fetus;
     - transfers carbon dioxide from fetus to mother;

- transfers wastes (or named wastes) from fetus to mother;
- allows the transfer of antibodies from mother to fetus;
- prevents mixing of the blood of mother and fetus.

c Helps prevent bacteria passing from mother to fetus.
The blood groups of mother and fetus may be different.

3 a (i) Sperm duct labelled correctly between the testes and urethra.
  (ii) Urethra labelled correctly between bladder and tip of penis.

b In males, the urethra carries urine and semen at different times; in females the urethra carries only urine.

c (i) Two male secondary sexual characteristics from:
  - voice becomes much lower (breaks);
  - hair starts to grow on chest, face, under arms and pubic area;
  - body becomes more muscular;
  - penis becomes larger;
  - testes start to produce sperm.
  (ii) Testis (or testes) labelled correctly.
  (iii) Testosterone makes muscles grow, so the athletes can run faster or perform better.

4 a Two points from:
  - using an unsterilised, used needle (in drug-taking or blood sampling);
  - unprotected sex with an infected person;
  - receiving a blood transfusion containing infected blood;
  - having a tattoo or body piercing using an unsterilised, used needle.

b Two harmful materials from:
  - drugs such as heroin;
  - paracetamol or aspirin;
  - nicotine;
  - alcohol.

Chapter 17

1 a

Type of cell	Number of chromosomes
Ciliated cell in windpipe	46
Red blood cell	0 (this cell has no nucleus)
Ovum	23

b Two differences from:
  - chromosomes in daughter mitotic cells will be identical to parental chromosomes (or there is no variation);
  - genes in daughter mitotic cells will be identical to parental genes;
  - chromosomes in daughter mitotic cells will be in homologous pairs, but they will be single in meiotic nuclei.

2 a Rr.
  b rr.
  c RR.
3 Gene, meiosis, diploid, phenotype, recessive, heterozygous.
4 A cross between two pink-flowering plants. The genotype of both plants must be C^RC^W phenotypes of parents.

phenotypes of parents	**pink**		**pink**
genotypes of parents	C^RC^W	×	C^RC^W
gametes	C^R C^W	×	C^R C^W

Punnett square

	C^R	C^W
C^R	C^RC^R	C^RC^W
C^W	C^RC^W	C^WC^W

F_1 genotypes	$1C^RC^R$, $2 C^RC^W$, $1 C^WC^W$,
F_1 phenotypes	**red** **pink** **white**
Ratio	1 red 2 pink 1 white

Chapter 18

1 a Selective breeding is a process used by humans. Individual plants or animals with desirable features are selected and crossed, and the most desirable individuals in the next generation are selected.

b e.g. Tomato plants bred for large fruit. The largest fruits are picked and their seeds are collected and planted. When the fruits are ripe, the seeds from the largest fruits are removed and planted. The process is repeated over a number of generations. Eventually a true-breeding variety of large-fruited tomato plants is produced.

Chapter 19

1 a A *producer* is an organism that makes its own food using energy from sunlight through the process of photosynthesis; a *consumer* is an organism that obtains its food by feeding on other organisms.

b A *carnivore* is an animal that eats other animals; a *herbivore* is an animal that eats plants.

2 a Two organisms in each column from those given in the following table:

Consumer	Producer	Carnivore	Herbivore
Bank voles, goldfinches, aphids, caterpillars, wrens, kestrels	Grass, hogweed, ivy, oak tree	Wrens, kestrels	Bank voles, goldfinches, aphids, caterpillars

b Six points, such as:
  - decrease in aphids because ladybirds eat aphids;
  - increase in ivy because there will be fewer aphids feeding;
  - decrease in wrens because there are fewer aphids to eat;

- decrease in caterpillars because the wrens now have only caterpillars for food;
- increase in oak trees because there will be fewer aphids feeding;
- increase in hogweed because there will be fewer aphids feeding;
- increase in goldfinches because there is more hogweed to eat.

Note that there are other possible suggestions.

3 a (i)   Light (or solar) energy.
  (ii) Chemical energy.
  b (i)   1200 units.
  (ii) 48 units.
  c (i)   The consumer may be warm-blooded, so some energy is lost as heat.
          Consumers usually move around to find food, find a mate or escape from predators, which uses up energy, but producers do not move.
  (ii) Feeding as a first consumer involves eating plants. Less energy is lost to the environment when feeding at this level, so food production is more efficient in terms of energy conservation.
4 a A, carbon compounds in plants; B, carbon compounds in dead plants and animals.
  b (i)   C on arrow between 'carbon compounds in fossil fuels' and 'carbon dioxide in the air'.
  (ii) D on arrow between box B and 'carbon dioxide in the air'.
  (iii) P on arrow between 'carbon dioxide in the air' and box A.
  (iv) R on arrow between 'carbon compounds in animals' (or box A) and 'carbon dioxide in the air'.
5 a (i)   Log phase (or exponential phase).
  (ii) The new population is taking time to settle or mature, before breeding begins. When this happens, a doubling of small numbers does not have a big impact on the total population size, so the line of the graph rises only slowly with time.
  b Three factors from:
    - amount of food available;
    - disease;
    - space;
    - predation.

Chapter 20

1 Two reasons from:
  - bacteria have a rapid reproduction rate;
  - can be grown and manipulated without raising ethical concerns;
  - have the ability to make complex molecules.

Chapter 21

1 a (i)   B.
  (ii) A.
  (iii) Bar for F drawn to 550. Column shaded in the same way as the others, and labelled. Column drawn an equal width and distance from the others.
  b (i)   30 years.
  (ii) Two reasons from:
    - to clear land for agriculture, housing, industry or roads;
    - to collect timber for housing;
    - to collect timber for fuel.
  (iii) Two suggestions from:
    - plants have gone, so there are no roots to bind the soil;
    - wind blows soil away;
    - rain washes soil away.
  c Increased carbon dioxide can lead to global warming, flooding or desertification.
    Organisms need oxygen for respiration to release energy.
2 a (i)   Constant level between W and X, or starts off quite low, or at point X it starts to drop then increases towards Y before dropping again towards Z.
  (ii) Level starts off quite low, then at point X it increases sharply; level returns nearly to original level between Y and Z.
  b (i)   One suggestion from:
    - aerobic respiration by sewage fungus;
    - lack of algae to produce oxygen.
  (ii) One suggestion from:
    - lack of sewage fungus;
    - photosynthesis by algae.
  c (i)   Two suggestions from:
    - presence of suspended solids blocks light for algae;
    - lack of nitrate in the water;
    - possible presence of toxins in sewage;
    - possible increase in temperature or unsuitable temperature.
  (ii) Two suggestions from:
    - shortage of nitrates;
    - grazing by aquatic herbivores;
    - possible drop in temperature, or unsuitable temperature.
  d One suggestion from:
    - herbicides could leach into river and kill algae;
    - herbicides will kill algae and disrupt food chains;
    - herbicides may be toxic to other organisms in the river.
3 a Combustion of fossil fuels.
  b (i)   The leaves are unable to photosynthesise, so it cannot make food.

Cambridge IGCSE Biology Study and Revision Guide Second Edition   © Dave Hayward 2016

(ii) One suggestion and explanation from:
- The roots are unable to absorb water, so cells will become flaccid, or the tree will wilt, or transport of materials will not happen.
- The roots are unable to absorb mineral salts that are needed, e.g. for formation of chlorophyll or for growth.

(iii) One suggestion and explanation from:
- Less decomposition will occur, so there will be fewer minerals available to the plant, e.g. magnesium ions for the formation of chlorophyll.
- There will be no nitrogen-fixing bacteria, so there will be fewer nitrates for the roots to take up, which are needed for protein formation.

Index

Cambridge IGCSE Biology Study and Revision Guide Second Edition © Dave Hayward 2016

It is illegal to photocopy this page

Cambridge IGCSE Biology Study and Revision Guide Second Edition © Dave Hayward 2016

Cambridge IGCSE Biology Study and Revision Guide Second Edition © Dave Hayward 2016